The
Blunder
Book

The Blunder Book

Colossal Errors,
Minor Mistakes, and
Surprising Slipups
That Have Changed the
Course of History

by

M. HIRSH GOLDBERG

Illustrations by Ray Driver

Quill/William Morrow/New York

Library of Congress Cataloging in Publication Data
Goldberg, M. Hirsh.
  The blunder book.

  Includes index.
  1. Errors, Popular. I. Title.
AZ999.G6 1984   001.9′6   84-4658
ISBN 0-688-07757-9 (pbk.)

Printed in the United States of America

First Quill Edition

1 2 3 4 5 6 7 8 9 10

BOOK DESIGN BY JAMES UDELL

*This Book Is Dedicated
with Much Love
to My Parents,
Herman and Ida Goldberg,
Who Have Assured Me
That I Was No Mistake*

If anything can go wrong, it will.

　　—Murphy's Law

Murphy was an optimist.

　　—O'Toole's Commentary

Contents

THE HALL OF FAME OF BLUNDER

1. Christopher Columbus (his error opened a whole new world)

2. The *Titanic* (the unsinkable design became the unthinkable accident)

3. The Leaning Tower of Pisa (a true monument to error)

4. The Edsel (a $250 million mistake—Detroit's first major flop)

5. The Liberty Bell (a cracked bell becomes the symbol of a nation)

6. Murphy (he laid down the law about error)

7. Ptolemy (his mistaken theory of the universe dominated astronomy for fourteen centuries)

8. Three Mile Island (a $4 billion accident and the first major mistake of the nuclear age)

9. Sigmund Freud (his Freudian slips gave a scientific name to errors of the tongue and pen)

10. The inventor of the eraser (for understanding the human condition)

INTRODUCTION

Oops:
The Reign of Error

The last words of General John Sedgwick,
uttered while peering at enemy lines during
the Civil War Battle of Spotsylvania
Courthouse in 1864: "They couldn't hit an
elephant at this dist—"

"To err is human, to forgive divine."
"Nobody's perfect."
"We all make mistakes; that's why pencils have erasers."

These are just some of the many expressions that recognize one of the continuing predicaments of the human condition—error. Indeed, in a world of humans striving to perfect themselves, life goes on very much imperfectly.

But how deep and pervasive is the imperfectness of our world? How much does error reign over our actions?

This book presents the many ways in which error has become ingrained in our government, science, medicine, and the arts, how error has changed history, crept into our libraries, altered our thoughts, and affected our daily existence. Here is shown the light side of error—the

13

amusing, surprising, unpredictable world that has resulted from accidents and mistakes. But shown, too, is the dark side—how flawed actions and faulty concepts have led to mishap and mayhem.

Indeed, error is so much a part of the human condition that our very mortality may be caused by a genetic code at the mercy of error.

In a 1981 article on the latest findings about the aging process, the *New York Times* reported that "one theory lays much of the blame for aging on an accumulation of errors in the master chemical of heredity, the nucleic acid known universally as DNA." As a result, to prolong life, some scientists are now "experimenting with ways of improving the body's ability to repair errors in the DNA."

Plant, animal, and human life in our present world may be both the survivors of error and its next victims. Paleontologists estimate that today's living species constitute but 5 percent of all the millions of species that have ever existed on earth. Although massive, unforeseen disasters may account for the extinction of numerous species, many others were those that somewhere along the ages made mistakes in adapting to changing conditions. Since conditions are always changing, some of the species now alive may eventually commit their own biologic blunders.

Although error may be an essential part of our lives, it is a part that in the past we have either ignored or endured—sometimes with humor but usually without further thought.

But the pantheon of error makers is large and always growing. As will be shown in these pages, history is replete with notables who have made serious mistakes, who have every right to hide their red faces from the limelight: people like Columbus, Gutenberg, Washington, Einstein. All of them—and many others—have had to overcome the errors of their ways to make their marks in the world. Sometimes, though, the errors overcame them. Johann Gutenberg, for instance, committed so many business blunders he lost his printing business and was not even around when the Bible that now bears his name was published.

And then there are the momentous events that have been shaped by error. The atomic bomb may have been dropped on Hiroshima because of a translator's slipup. Germany committed a strategic mistake—and lost World War I—because of a blunder by its foreign minister. The celebrated Charge of the Light Brigade was really the murderous result of military snafus.

As much as we would like to believe to the contrary, the actions of our most esteemed institutions and the pronouncements of our most respected experts are fraught with error. A physicist checking out the venerable *Encyclopaedia Britannica* discovers numerous erroneous statements and, taking one of its recent editions as a case study, finds more than 600 articles with outdated or discredited information. The *New York Times* editorializes earlier this century that space rockets will never be successful. The British government, back in the 1960s during a time of fiscal gloom and consumer austerity, discovers it has been making a $1.6 billion accounting error—against itself.

As for the advice of experts, during 1981 in a major case against the Ford Motor Company, the plaintiff's lawyer, a renowned attorney, in hopes of an even bigger judgment, tells his client to forgo Ford's offer of a $2 million settlement made while the jury is out deliberating the case. The jury soon returns to the courtroom—and finds Ford innocent of all charges. The plaintiff is now suing his lawyer.

Errors abound, both in private and in public life. The world appears to lurch forward by accident. Actions seem to be taken without sufficient planning, other times with planning that is foolish and flawed. Too many times does an event fall apart because of somebody's foul-up.

Error is not always bad, and it sometimes leads to productive results. Penicillin, X rays, rubber, photography, electric current, and the telescope all were discovered by error. Even one of our foods was created by mistake. The Belgian endive, described by food critic James Beard as "one of the most delectable and delicate vegetables in the Western world," was cultivated in 1843 when a gardener in Belgium accidentally left chicory roots in his moist, dark cellar; when he came back a few days later, he found a never-before-seen vegetable growing. Deprived of sunshine, the roots had sprouted delicate white leaves that proved tasty. Beard has termed the "absolutely delicious" Belgian endive "a magical vegetable."

Also, humanity's basic way of achieving advances has been through trial and error. Thomas Alva Edison thought nothing of going through thousands of unsuccessful experiments to arrive at a solution for an invention. This use of trial and error prompted his famous statement that genius is "one percent inspiration and ninety-nine percent perspiration." A magazine advertisement for a commodities trading house stated in a headline beneath a picture of Edison, "He who is afraid to make mistakes is afraid to succeed."

Maybe because of this, and because they are so ubiquitous, we tend to have a soft spot for errors and those who make them. The first comedy album to sell a million records was *Radio Bloopers,* issued in

1954. We have made best sellers out of such books as *Murphy's Law,* a compendium of sayings that touch on the general propensity—no matter how hard we try to avoid them—to make mistakes of our own and be forced to endure the foul-ups of others.

The work before you is a fleshing out of Murphy's Law, a presentation of actual incidents, true stories, facts, profiles, and statistics that show how prevalent is humanity's tendency to err and how we must always be on guard against error in what we do—and what others do to us. Here are the people and events to prove not only that "if anything can go wrong, it will" but also that "everything can go wrong, and does." What follows in these chapters is a reminder that error will always be with us, that it will always have its star performers (Wrong Way Corrigan), its philosophers (Murphy for his Law and Freud for his Freudian slips), its products (the Edsel), and its symbols (the Leaning Tower of Pisa).

In short, this book illustrates that our battle with mistake and mishap is a continuing struggle—and a costly one. It also shows that error is but the counterpart of our pride, that no matter how hard we try, we do not have the ability or the power to control fully our destiny. For better or for worse, error is evidence that we are human.

Yes, to err is indeed human—and when we err, to be forgiven is just divine.

If Lincoln were alive today, he'd roll over in his grave.

—President Gerald Ford

CHAPTER I

Blunder: Error in History

"Do you ever admit a mistake?" President John Kennedy asked Soviet leader Nikita Khrushchev in 1961 in Vienna during foreign policy talks.

"Certainly," Khrushchev replied. "In a speech before the Twentieth Party Congress, I admitted all of Stalin's mistakes."

—*Presidential Anecdotes*
by Paul F. Boller, Jr.

We live in a universe the origin of which, according to many scientists, was an explosion. This is the big bang theory of how our world began.

We live out our days on the remains of that explosion, on a planet which rotates in space not straight but tilted at its axis, a globe which at one time was considered by everyone living on it to be flat and which today is widely believed to be round but isn't (it is slightly depressed at the top and bottom to create a shape officially called an oblate spheroid).

Is it any wonder, then, that the history that has been played out on this place called earth—residue of an explosion, tilted, squashed, and populated by DNA mutations—has been affected by error?

We may wish to believe that the great forces of history move without mistake, but the truth is far different. The events of history are more often than not characterized by bungling and blunder because people are the ones involved in making that history—kings and their subjects, presidents and private citizens, generals and soldiers, the high and mighty and the low and mighty poor. As a result, human error shapes and colors happenings as much as, if not more than, the careful planning that we would like to believe controls events.

Consider that one of the major instruments for creating history is war. The borders of our lands as well as the landscape of our minds would be different if the human being did not try to extend his sway over others by force—and had not often succeeded. But even with such success, wars generate an abundance of mistakes. Battles such as Waterloo and Custer's Last Stand are synonymous with failure and faux pas. The problem is that warfare has been the province of the military, and therein often lies the problem in history. As Groucho Marx once declared, military intelligence is a conflict in terms.

Indeed, the age of nuclear warfare may have opened by mistake.

On July 26, 1945, leaders of the United States and Great Britain issued the Potsdam Declaration, in which Japan was called upon to surrender unconditionally or face destruction. Japan's Emperor Hirohito inquired through the Russians if the reference to "unconditional" surrender could be deleted since on June 22 Japan's Supreme Council and the emperor had decided to secure a negotiated peace. The Russians informed President Harry Truman about Japan's request, but he refused to alter the demand for unconditional surrender. Japanese leaders then decided that instead of responding directly, they would wait to see if further diplomatic moves might bring about the negotiated peace they desired. A statement was issued to the world press announcing Japan's intention to forgo responding for the present to the unconditional surrender terms. In Japanese, the statement used the word *mokusatsu,* which has two possible meanings: (1) to ignore, and (2) to refrain from comment. But then the glaring error was made: The Japanese translator, in the English version, did not use the intended "refrain from comment," but instead wrote that Japan would "ignore" the demand for unconditional surrender.

When American leaders read the statement, they became incensed. They saw the Japanese intention to "ignore" the call for unconditional surrender as tantamount to expressing contempt for any proposals by the United States. Although the mistranslation was not the only cause of the increasing friction of the time, it spurred an escalation of tension between the United States and Japan. Less than two weeks later, on August 6, 1945, the United States dropped the atomic bomb on Hiroshima.

Historians and philosophers may debate whether people or events affect history more, but one thing is certain: Error affects the people and events of history more than we realize.

Here, in chronological order, is a look at some of the great blunders in history.

ONLY THE PINUPS ARE ACCURATE: THE WESTERN WORLD'S ERRONEOUS CALENDAR

Since the recording of history is tied into the dating of events, we begin our survey of historical blunders by looking at how one man's error has affected the dating of history for the past 1,500 years, a mistake that has made the calendars of the Western world incorrect by 4 to 20 years.

Since the sixth century the West—and much of the rest of the world—have numbered the years by citing "Year One" as the year in which Jesus was said to have been born and labeling all the other years in relation to it. This approach is based on the calculations of a monk—Dionysius Exiguus (Denis the Little)—who presented his research on Jesus' birth and his calendar ideas A.D. 525 (at least he said it was A.D. 525). However, Dionysius, who has been termed one of the most learned men of the sixth century, made a serious error which, at that time, went unnoticed and which, since then, has gone uncorrected.

Prior to Dionysius, the Western world followed the Roman calendar, which began with the year in which Rome was said to have been founded. That year is now referred to as 753 B.C., but at that time the year was called 1 A.U.C. A.U.C. is the abbreviation for the Latin *anno urbis conditae,* which means "the year of the establishment of the city."

When Dionysius came along, he advocated renumbering the years, basing the designation not on the founding of Rome but on the birth of Jesus. The monk, however, made a mistake trying to determine the year. According to *The New Catholic Encyclopedia,* Dionysius "wrongly dated [the birth of Jesus] to 754 A.U.C., some 4 years, at least, too late." Thus, what Dionysius said was A.D. 1 should have been four or more years earlier.

19

While scholars now reject Dionysius's reckoning, there is still little unanimity about what year to cite as the one in which Jesus was born. Various authorities on the Bible have placed the birth as far back as 20 B.C.—and even as far forward as A.D. 6. One major basis for figuring the birth is the reference in the New Testament that Jesus was born during a census taken by King Herod. We now know that Herod died in 4 B.C. and the census referred to may have taken place up to three years before that. Thus, the most accepted view today is that Jesus was born toward the end of Herod's reign. *The New Catholic Encyclopedia* states that "the most probable date seems to be about the year 7 or 6 B.C."

One thing, though, is clear: Jesus was not born A.D. 1, and all the calendars, history books, and reference works that use this dating system are incorrect. And all because of the error made by a monk A.D. 525, which was really A.D. 521 or A.D. 519 or A.D. 518 or . . .

WHEN HISTORY LOST TEN DAYS

Humanity seems to have had great difficulty in getting the calendar to work properly so that year in and year out the changing seasons are accounted for and the holidays and other special days are commemorated annually in much the same setting and time of year. Numerous civilizations have tried out their own calendars. Some, such as the Jewish calendar, are still followed and still workable after thousands of years. But others are not so reliable and have to be revamped after several centuries.

The Roman calendar, called the Julian calendar after Julius Caesar, was one such example. The Romans may have been accurate in building aqueducts and coliseums, but they erred when it came to the calendar. They started out with ten months, then added two more months, while fiddling with the number of days in February (at one time it had 30 days). All this was all right, but they figured the year to have 365¼ days. After a period of time they noticed that the months were starting to slide out of sync with the seasons. What they did not realize was that the year really contains slightly less than 365¼ days—actually eleven minutes and fourteen seconds less than that. By A.D. 730, when this mistake was spotted by a monk (those monks were real calendar watchers), the calendar was 5½ days ahead of the seasons.

The monk tried to get adjustments made, but nobody listened—until 800 years later.

The Julian calendar was finally reformed in 1582, when Pope Gregory XIII, noting how Easter was coming later every year, proposed a series of changes to correct the error and keep the seasons and the calendar in line. He shortened September 1582 by ten days and decreed one out of every four centuries should begin with a leap year. This would serve to fine-tune the calendar by subtracting three days from the calendar every 400 years. But since the Gregorian calendar, as it was called, necessitated the elimination of ten days from the calendar then in use throughout much of the Western world, a number of countries balked.

Another reason for the reluctance was religion. The ones that immediately went along with the plan were the Catholic countries in Europe, but most of the Protestant countries refused. Also, nations that followed the Greek or Russian Orthodox liturgy—such as Greece, Serbia, Bulgaria, Rumania, and Russia—resisted, as did Moslem countries, which did not accept either the Julian or Gregorian calendar.

Slowly but surely, however, over the years even countries not controlled by Catholics agreed to use the Gregorian calendar since the incorrectly calculated Julian calendar was now throwing the year clearly out of whack with the earth's rotation. But some changeovers were long in coming (Russia did not change until 1918).

One of the nations that delayed converting to the new calendar was colonial America. Not until 1752 did England and its colonies finally drop the Julian calendar and adopt the Gregorian—nearly 200 years after it had been proposed. And even then the change caused many Englishmen to go berserk because they believed reforming the calendar had shortened their lives. "Give us back our days," the mobs cried out.

The changeover in America led to the loss of eleven days in February 1752 to adjust to the Gregorian calendar. This affected, among others, George Washington, who as a twenty-year-old in 1752 saw his birthday moved from February 11 to February 22. Washington himself continued to celebrate his birthday on the eleventh, and the national celebration of his birthday first took place on this date. Not until 1796, the year he left the presidency, did America begin observing Washington's birthday on February 22. (Of course, the celebration of George Washington's birthday has now been shifted a third time—to the third

21

Monday in February—which has nothing to do with the Julian or Gregorian calendars, but with an even more important one for most people: the federal holiday calendar.)

THE ADMIRAL OF ERROR DISCOVERS THE NEW WORLD

Christopher Columbus, the discoverer of the New World, the man who insisted on being called the Admiral of the Ocean, had a life bounded, rescued, and finally made famous by error.

We ourselves generally are in error about his accomplishments. Christopher Columbus (1451–1506) discovered the New World not because, as generally believed, he was alone in realizing the world is round. This was a concept first advanced by the ancient Greeks and was well known and largely accepted by the intelligentsia and upper class of his day.

What Columbus advocated—and had trouble convincing others of —was that the earth was 25 percent smaller than generally believed; to find a shorter route to India, he said, he could sail westward with the ships then available and still be able to reach India.

He also said the size of the Asian continent was larger than estimated, believing that the farther it extended toward the east, the nearer it came toward Spain and, therefore, that much easier for him to reach by going to the west.

What was at issue, then, was the size and scope, not the shape, of the world.

The truth is that Columbus was wrong and the leading scientists and geographers of the day were right. The planet was not as small, and Asia not as big, as Columbus had proposed.

What saved Columbus from his miscalculations was that he bumped into the unknown landmass that stood between Europe and India.

Columbus, though, did have to overcome other errors in the thinking of his day. In 1490 King Ferdinand and Queen Isabella of Spain appointed a royal committee to study Columbus's proposals to sail westward. The consensus of the committee was that the project was impossible. Among the reasons given:

1. Duration—such a voyage would take three years.
2. Obstacles—the ocean was seen as infinite and possibly unnavigable.

3. Difficulty of return—if he reached the land on the other side of Europe (called the antipodes), Columbus would not be able to get back.
4. Lack of landing room—water covers most of the earth; besides, St. Augustine had said there was no antipodes.
5. The objective—it was unlikely that so long after the world had been created any unknown lands of value could be found.

It was this constricted thinking that Columbus had to overcome. He finally prevailed over his opponents and received the approval of the Spanish monarchy for the voyage. His errors had triumphed over other people's errors.

But when Columbus bumped into the New World, his penchant for error continued. First, he thought he had indeed landed in India and dubbed his discovery West Indies and the natives Indians. Later he discovered what is now Costa Rica and gave it that name, which means "rich coast," because he saw natives with gold necklaces and believed the land was rich with gold and silver. However, Costa Rica has proved to be among the Latin American countries with the *least* amount of mineral wealth. But as with the West Indies, the name Costa Rica stuck, remaining on the map as another example of Columbus's errors.

For all his naming of places in the New World, Columbus missed out not only on naming the most important land of all—the continents themselves—but on having them named in his honor. That honor went to a little-known Italian merchant—explorer Amerigo Vespucci, who claimed to have discovered the continent of America in 1497. Since Columbus at that time still had no idea he had reached the Western Hemisphere but continued to believe he was exploring parts of the Indies (he did not step onto the mainland of South America until 1498), he did not dispute Vespucci's claim. Vespucci went on to publish letters about his discovery, and in 1507, when a mapmaker needed a name for the new continent (which was South America), he suggested America "because Amerigo discovered it."

Scholars no longer believe Amerigo was the discoverer of South America, and his name in association with it is another error of history, but the name America was accepted first for South and then for North America. Today, however, the continents of the New World might be called North and South Columbus if the Admiral of the Ocean had not made his biggest error of all.

THE BUBONIC PLAGUE

Known in its most virulent form as the Black Death, the bubonic plague raged off and on for three centuries during the Middle Ages, eradicating a fourth of Europe's population, killing an estimated 25 million people, and proving to be one of the greatest catastrophes of nature to afflict the human race. But one major reason it was able to wreak such havoc was people's tendency to resort to outlandish, foolhardy remedies, some of which were fatal themselves.

The plague is believed to have originated in China, then eventually was carried by merchant ships to Italy, where in 1348 diseased rats came ashore in Genoa and began infecting the populace almost immediately. The symptoms of the disease were obvious and often fatal: headaches, followed by fever, shivering, and then dizziness, after which black, hard boils erupted in the groin area and in the armpits, the lymph nodes swelled, and the victim began vomiting blood. In three days the person died.

The medicine of the day was powerless to stop the spread of the Black Death, which at its worst was a combination of bubonic plague, typhus, and cholera. Alternating in intensity over the centuries, it eventually afflicted 200,000 towns and villages in Europe, killed more than 1 million people in one year in Germany, decimated the populations of Iceland and Cyprus, destroyed half the population of Italy, infected 9 out of 10 people in London.

Compounding the problem of disease, however, were the remedies devised either to halt the spread of the plague or to cure its victims. These ideas read like a litany of lunacy. Afflicted persons were told to drink melted gold and ground emeralds; swallow menstrual blood; spread lard into open wounds; pierce testicles; press to feverish foreheads the blood from pigeons or one month-old puppies; cut out boils, dry them, and make them into a powder which was then to be eaten.

The preventive measures were built around the weird concept that the Black Death could be kept away if the air were made to smell bad enough. This was the theory of stinks. German and Balkan people were fanatic believers that stinks would fumigate the air and prevent the plague.

To create these stinks, it was advocated that dead dogs be thrown into the streets and that in the bedroom one should keep a live, stinking billy goat. One doctor called for the wearing of human excrement; another recommended "bottled wind."

While one monk suggested goat urine as a good stink, a thinker of the day declared, "A wash with urine does more than any other preventive, more particularly when in addition, the urine was drunk."

The plague became less of a major health problem toward the end of the seventeenth century, but without apparent reason at the time. Looking back, we can see a possible cause. After accusing Jews and then dogs of spreading the disease, people began to realize that fleas from diseased rats were the culprit. Also, public sanitation and private hygiene had begun to be widely practiced. As one history of the Black Death speculates, simple soap and water—mixed with common sense —had finally stopped the germ.

THE PILGRIMS LANDED IN THE WRONG PLACE

It may confer great social status to have an ancestor who came to America aboard the *Mayflower,* but the first Pilgrims landed in the wrong place because of a navigational error on the *Mayflower.*

On September 6, 1620, (September 16, New Style), the Pilgrims sailed from England, heading for the Atlantic coast of North America near the Hudson River. They had been granted land in that area on which to build their new colony and start a life of religious freedom. Instead, because of a mistake in navigation, on November 10/20, the *Mayflower* arrived north of the site, off Cape Cod. The Pilgrims turned the *Mayflower* to try to right their course, but within hours they encountered heavy breakers and dangerous shoal water. The next morning, a Saturday, November 11/21, found the *Mayflower* back at the area off Cape Cod, where, as the sun came up, the Pilgrims decided to drop anchor.

The first Pilgrims who later that day set foot on land did so knowing they had come ashore in the wrong place. But since they were now without laws (they were outside the North American jurisdiction of the company that had sponsored them), even before they left their ship they drew up what became known as the Mayflower Compact. The compact, the first agreement for self-government ever put in force in America, was virtually their constitution, a body of laws for their settlement in North America and the establishment of a government in what became Plymouth Colony. Thanks to their landing error, the basis for the Pilgrims' government was not that of the charter company but their own desires. The Mayflower Compact was a significant beginning for a New World rather than a carry-over of the old.*

THE DECLARATION OF INDEPENDENCE

"When in the Course of human events," the text starts. One of the great documents in the history of the world. The precursor not only of the American Revolution but later of the French Revolution and other revolutions that have swept away rulers who would deny freedom to their people.

But the creation of the Declaration of Independence did not happen as neatly as we would expect. Indeed, untidy mishaps and mix-ups attended the birth of a document that has always seemed to have emerged full-blown, tied into a red, white, and blue bow.

First, the document begins, "In Congress, July 4, 1776. The unanimous Declaration of the thirteen united States of America." About the only thing true here is the reference to the year. The implication that every state agreed to the Declaration and that it was signed by Congress on July 4 is not true. It was on July 2, 1776, by a vote of twelve states (with New York not voting at all and two other states—Pennsylvania and Delaware—casting divided ballots) that Congress adopted the resolution of independence. On the fourth came Jefferson's Declaration, but it was the second that saw the decisive act of independence. Thus, for those who are sticklers, the July 4 celebrations today are in error by two days.

Actually July 4 marks little of significance. New York did not vote its acceptance of the Declaration until July 9, and that vote was not announced to the Congress until the fifteenth. But it was not until the nineteenth that the Declaration, finally being unanimous, was ordered engrossed and signed. The lack of unanimity until then can be seen in the fact that when it was read to the army during this period the

*One episode about the *Mayflower* bears telling for what it says about those who would flirt with error. The nine and a half weeks of the voyage were marked by storms and squalls, and many of the passengers became seasick. A young crew member, however, made repeated fun of the ill, mocking them and predicting half of them would not survive the trip, in which case, he told them, he would gladly bury them at sea and help himself to their possessions. Rebukes by those on board could not stop his laughing and jeering. Shortly before land was sighted, the only death among the passengers occurred, but the offensive sailor did not see this. Before the *Mayflower* reached the halfway point, the jokester himself became ill and died. His was the only fatality among the crew. [See *The Mayflower* (Stein and Day, 1974) by Kate Caffrey] opening was worded: "In Congress, July 4, 1776. A Declaration by the Representatives of the United States of America, in General Congress assembled." A meaningful difference, one that in a sense was hushed up by the switch to the wording that has always been associated with the document.

A great deal has been made about all the signatures accompanying the Declaration of Independence. In fact, John Hancock probably made a career out of his florid signature (a life insurance company surely has). Although the document implies that all the signatures were applied on the fourth, in fact most of the signers of the Declaration affixed their names on the document on August 2. At least six signed later. In addition, not all who voted for the Declaration ever signed it. And not all who signed it were members of Congress when the Declaration was adopted. Finally, one man—Thomas McKean, who was present on July 4—did not sign his name until 1781.

As for the document's great declarations about freedom from oppression, Thomas Jefferson's noble denunciations of slavery were dropped from the final draft.

Thus, little about our perception of how the Declaration of Independence came into being is correct.

After the Declaration had passed into American history, the document itself was handled incorrectly. For 101 years after passage, it was shunted around to ten cities in five states, was exposed unprotected to light for 50 years, and was frequently rolled and unrolled, thereby damaging the signatures. In 1921 the Declaration of Independence was finally placed in the Library of Congress, and it now rests on exhibit in Washington, D.C., in a specially sealed showcase. But the damage was done. Today the writing is so faint as to be hardly readable, and estimates are that the decay is irreversible. Because of the years of mishandling, within several decades there will be nothing left to read on the original document. While other older documents (such as the Magna Carta and the Dead Sea Scrolls), have survived, the Declaration of Independence will soon be just a blank piece of parchment.

GEORGE WASHINGTON, FATHER OF BLUNDER

American history might have been far different if not for George Washington, but George owed much of his success to his mistakes.

George Washington has a special place in American history in part,

REVOLUTIONARY MATHEMATICS

In the room where the Declaration of Independence is now kept—a special exhibit area of the National Archives in Washington, D.C.—are several other documents from that important era in American history. One of these—called Lee's Resolution—shows the tallies taken of an important vote on July 2, 1776, involving the affirmative votes of twelve states. On a corner of the document is a handwritten list of numbers, but they are added incorrectly.

The list of numbers and the sum given for them are:

$$90$$
$$81$$
$$96$$
$$\underline{96}$$
$$383$$

The correct answer is not 383 but 363.

What effect on the voting that day—and on American history—did this error have? No one is certain. In the National Archives exhibit in which this document now rests is this note: "The incorrectly added figures in the lower right are a mystery."

---

ironically, because the man who could not tell a lie had many lies told about him. One such fabrication is the story of how he supposedly chopped down a cherry tree but admitted his terrible deed to his father. This tale was invented by Mason Locke Weems, better known as Parson Weems. Weems, born in Maryland in 1759 and the main source of writings about Washington during his lifetime, was quite a character. He did not hesitate to play on people's credulity or stretch the facts. On the title page of his Washington biography, in which the story of the cherry tree was first presented, he identified himself as "Formerly Rector of Mount Vernon Parish." No such parish ever existed.

But if Washington's initial biographer and biography are flawed, so, too, was our hero. Father George, it seems, was not the stunning field commander and general that American legend says he was. The *Ency-*

29

clopaedia Britannica states he was "guilty of grave military blunders."
Michael Korda, in *Success!* (Random House, 1977), details how Washington was a case in point of "upward failure." During his lifetime he had a reputation for losing battles ineptly. Thomas Jefferson said of him
—delicately, it must be added—that he was "not a great tactician" and that he "failed in the field." John Adams was less reserved, terming him
"an old muttonhead."

Among his military blunders, George, then a young officer, chose a swampy area constricted on three sides by tree-filled hills as the site for a fort to be built at the Great Meadows, Pennsylvania. The location proved so strategically poor that the enemy, at that time the French, easily captured Washington.

After being freed by the French, Washington went on to become personal assistant to Lieutenant General Edward Braddock. Soon Washington was advising Braddock on how to handle the forces at his command during the siege of Fort Duquesne. What happened is something Parson Weems surely did not discuss. The army was ambushed, and General Braddock lost not only the battle but himself. Writes Korda: "Washington, whose advice had been so conspicuously wrong, went on to blunder in larger matters."

Washington's higher military muffs began when the Continental Congress made him in 1775 commander in chief of the Continental Army. By 1776, for instance, he was predicting that victory in the War of Independence "depends in all human probability on the exertion of a few weeks." The war, however, took seven more years.

George Washington may have caused his own death. On December 12, 1799, after writing a letter to Alexander Hamilton, he went for his daily horseback ride around Mount Vernon, even though the weather was cold, with snow turning into sleet and rain. When he came back after five hours, he decided not to change his damp clothes but sat in them through dinner. The next day he awoke with a sore throat, and even though it was snowing, with three inches of snow on the ground, he went for a long walk. Early the next morning, December 14, 1799, Washington awoke, feeling ill and suffering from laryngitis. But he would not let his wife, Martha, send for a doctor until dawn. During the course of that day, with three doctors finally in attendance, Washington was bled four times. But nothing could stem his decline. By 10:00 P.M. he had whispered his last—" 'Tis well"—and died.

George Washington, first in the hearts of his countrymen, had blundered into his last moments.*

AND REMEMBER—HE NEVER LIED

George Washington, in his Farewell Address as president, declared: "Though in reviewing the incidents of my Administration I am unconscious of intentional error, I am nevertheless too sensible of my defects not to think it probable that I may have committed many errors."

THE LIBERTY BELL: SYMBOL WITH A FLAW

The Liberty Bell, according to legend, was rung on July 4, 1776, to announce the adoption of the Declaration of Independence. Pictured on stamps and coins, the Liberty Bell is often cited as a symbol of American freedom, for engraved on it is the saying from the Bible (Leviticus 25:10) "Proclaim liberty throughout all the land unto all the inhabitants thereof."

The Liberty Bell, however, was so poorly cast that it broke three times while being rung and developed a gaping crack down its front which has made it virtually useless as a bell. Today the crack is highly visible on this flawed national symbol.

Commissioned in 1751 by the Pennsylvania Provincial Assembly to hang in what is now Independence Hall in Philadelphia, the bell was delivered in August 1752, after being cast in London, England, by the Whitechapel Bell Foundry.

The bell cracked for the first time while being tested. It was then twice recast in Philadelphia before being hung in the Statehouse steeple in June 1753. It broke again in 1835 while tolling during the funeral of U.S. Chief Justice John Marshall. In 1846, while being rung for George Washington's birthday, the bell cracked a third time, this time so badly that it could not be repaired.

The Liberty Bell, which is no longer rung, rests today in Philadel-

---

*George offered an interesting variation of the phrase "To err is human." After a violent argument with a William Payne about a contest for the Virginia Assembly, Washington apologized, saying, "To err is nature, to rectify error is glory."

31

phia, revered as a major element in American history. It is a strange symbol for a great nation, however. The legend that it was rung to signal the Declaration of Independence is now considered to be erroneous. And thanks to errors in casting, America has a cracked, improperly constructed, and unusable bell for a national symbol.

THE ROYAL FAUX PAS THAT HELPED LAUNCH THE REIGN OF TERROR

The Reign of Terror was a period of time during the French Revolution in which 20,000 people were executed by revolutionaries intent on wiping out vestiges of the monarchy and any opposition to the Revolution. The execution of Louis XVI (1754–1793) in January 1793 marked the beginnings of the headlong rush into the Reign of Terror.

Louis XVI had desperately tried to avoid his fate, but his own faux pas did him in. Since October 1789, he had been held by revolutionaries as a virtual prisoner in the Tuileries Palace. On June 20, 1791, accompanied by his beautiful wife, Marie Antoinette, and their children, Louis attempted to escape from Paris and reach the safety of what is now the Belgian border. Masquerading as a middle-class merchant, he dressed inconspicuously in a brown wig and a plain gray overcoat and quietly departed the city in a coach similar in appearance to others of the day. His three guards were disguised as coachman and footmen.

But Louis forgot that his picture as king was widely available—on the money of France. As the horses were being harnessed at the Golden Sun hostelry where his contingent had paused, he did not heed one of his guards who warned him to remain hidden, but leaned out the window of the coach to ask directions. In a royal case of Murphy's Law, Louis was observed by one of the few people in the area who could recognize him. The postmaster of the small town, a young man named Jean Baptiste Drouet, who had seen the king while serving in the cavalry stationed in Paris, noticed the brown-wigged face and thought it seemed familiar. When Drouet was paid by one of Louis's men and looked at the money, he realized that the face of the king on the money and the face of the man in the coach were the same.

As soon as the coach and the king departed, Drouet ran to the town hall, where he announced his discovery. Soon a contingent of local National Guard galloped after the fleeing king. At the little town of Varennes the group surrounded the coach and took the king and his

COMMONLY MADE MISTAKES ABOUT TWO HISTORIC DOCUMENTS

The Magna Carta, often cited as marking a decisive step in democracy, had nothing to do with democracy or with the rights of the masses. In 1215 the English barons forced King John to sign a charter (*Magna Carta* means "Great Charter") granting certain rights to the nobles and merchants in chartered cities. The rights of the common man in England were not addressed by the charter. In its own time, as *The World Book Encyclopedia* notes, "the greatest value of Magna Carta was that it placed the king under the law, and decisively checked royal power." That checking of royal power related solely to the privileged classes. Only in later centuries did the Magna Carta become a model for those demanding a democratic form of government for all and rights for the individual.

The Emancipation Proclamation, said to have ended slavery in the United States, did not free a single slave. The proclamation, signed into law by Abraham Lincoln in 1862 to take effect in January 1863, called for the freeing of slaves *only* in those states which did not return to the Union fold by the beginning of 1863. But those states which remained in the Confederacy were obviously not under Lincoln's control and did not give their slaves their freedom in response to Lincoln's proclamation. All other states did not have to free their slaves. In fact, the document specifically excluded the freeing of slaves in the border states and in those southern states then under Union control, such as Tennessee and parts of Louisiana and Virginia. The true purpose of the proclamation was to discourage Britain and France from fighting on the side of the Confederacy (the two countries relied on southern cotton, but they opposed slavery) and to help the North's war effort by enabling escaping blacks to join the Union army. It was not until after Lincoln's death and the ratification of the Thirteenth Amendment to the Constitution on December 18, 1865, that slavery was officially abolished throughout the United States. Until then the main purpose of the Emancipation Proclamation was not to gain freedom for slaves, but to force Southern states to return to the Union.

family into custody. They were returned to Paris, where the following year (August 10, 1792) Louis was removed as king and put into prison. He was eventually placed on trial and convicted of crimes against France. On January 21, 1793, instead of being alive and free in a nearby country, from which he might have restored the throne, Louis XVI was executed by guillotine.

While Louis's demise was preceded by a slipup, his birth was accompanied by an accident that proved symbolic of his fate. After his birth on August 23, 1754, a courier rushed by horse from Versailles to carry the news to the child's grandfather, King Louis XV. In his haste the messenger whipped the horse to such a frenzy that the animal reacted by throwing the rider, who flew from the horse and plunged onto the road, breaking his neck. Thus was heralded the birth of the king who would lose his head to the guillotine.*

THE WAR TO END ALL WARS

Our nomination for the War to End All Wars is not World War I, as it is often described, but the Crimean War, a war which will live in infamy—and in the annals of error. The Crimea is a peninsula of Russia surrounded by the Black Sea. From 1853 to 1856 it was the site of a bloody conflict which saw Britain and France joining forces. Along with Turkey and Sardinia, they fought Russia over various issues, ranging from dominion over southeastern Europe to a dispute about control of the holy places in Jerusalem. The results were eventually favorable to the allies, but the war has long been considered a mistake for everybody concerned. Today it is primarily remembered for the Charge of the Light Brigade—an insane, suicidal effort which caused more than 100 needless deaths because of a mix-up in orders

---

*Louis XVI's life seems to have been a combination of the Reign of Terror and the Reign of Error. For the first seven years of his marriage to Marie Antoinette he found himself incapable of consummating the marriage, let alone fathering an heir to the throne. He could not figure out what was the matter and for years suffered from his wife's gibes about his "problem." His impotency was eventually diagnosed as being due to phimosis (a painful narrowing of the foreskin), and he was finally cured in the summer of 1777 by a minor operation. Thereafter, according to historian Saul K. Padover in the first full-length biography in modern times of the Bourbon king, *The Life and Death of Louis XVI* (originally published in 1939 and reissued by Taplinger in 1963), "Louis was beside himself with joy and became an unbearably attentive husband."

yet is celebrated as an act of courage in a poem written soon thereafter.*

The mismanagement of this conflict can be seen in a delightful essay on the Crimean War in the fourteenth edition of the *Encyclopaedia Britannica*. Written by F. J. Hudleston, late librarian of the War Office in London, and E. W. Sheppard, director of recruiting and organization in the same London War Office, the article is filled with acerbic comments on this five-nation attempt at waging war. Some of the bonnier bon mots:

—The British commander of the war was Lord Raglan,† the Fitzroy Somerset of the Peninsular War, who had seen no service since 1815 and who had an "incurable habit throughout the campaign of referring to his enemy as 'the French.' "

—Britain and France knew very little about the Crimea. "The British cabinet, however, observing from a cursory glance of the map that the Crimea was a peninsula, conceived that there could be nothing easier than for the British fleet to cut it off from the mainland by commanding the isthmus with its guns—nor could there have been but for the fact, subsequently discovered, that the depth of the sea on either side of this isthmus was little more than two or three feet."

—The Russian governor of Evpatoria, near where the allies decided to land, "on receiving the formal summons to surrender . . . first fumigated the document, then read it, and, realizing that he must yield to superior numbers, insisted that the British and French troops on landing must consider themselves in strict quarantine." The allied troops landed without opposition.

—"The Russian army was ever behind the times. The regiment belonged to the colonel, not the colonel to the regiment. . . . its tactics were still based on

---

*The Crimean War is also celebrated for marking the first time in which Florence Nightingale performed nursing for injured soldiers. Until then the only women appearing near a battlefield were prostitutes or nuns.
†Yes, the raglan sleeve was named after Lord Raglan.

Suvorov's motto, 'The bullet's a fool, the bayonet's a fine boy.' "

—"As no combined plan of attack on the Russian position behind the Alma river had been arranged beforehand, cooperation between the Allies was conspicuous by its absence, and they fought two actions side by side. . . . Generalship was equally absent on the side of the Russians where 'no one received any orders and every man did what he thought best.' The steady advance of the British up the slope across the river made an unforgettable impression on the French general, Canrobert; they went forward, he said, 'as though they were in Hyde Park!' "

—"The Allies had no maps of the Crimea, and those in the possession of the Russians were so indifferent that one regiment, after marching steadily for the whole of the 20th, finally found itself back in front of Sevastopol [the city where they had started]."

—When Marshal St. Arnaud, commander of French troops, died, he was succeeded by Canrobert, who to the English "appeared with his gestures and grimaces like a play actor" and whose "serious defect was that he always thought so much was to be said on both sides of any question that he could never make up his mind which side had most to be said for it."

—When an electric cable which cut communication time between the field and national capitals from ten days to twenty-four hours was introduced, the French used it to send advice, instructions, and suggestions to their commander in chief, while the British War Office "concerned itself more with enquiries as to the health of Capt. Jarvis, believed to have been bitten by a centipede, and heated discussion as to whether beards were an aid to desertion." As a result, Raglan's successor, General Sir James Simpson, "was kept at work answering correspondence from 4 A.M. to 6 P.M. daily," causing him at one point to remark that "the telegraph had 'upset everything.' "

—The Crimean War was the one in which there occurred the Charge of the Light Brigade, termed here "one of

the most heroic and useless episodes in English military
history."
—Summing up, our authors state: "As far as concerns the
military art, the Crimean War is usually regarded as
worthy of remembrance only as perhaps the most
ill-managed campaign in English history."*

THE CHARGE OF
THE LIGHT-IN-THE-HEAD BRIGADE

The Crimean War was the occasion for what is widely regarded as one
of the stupidest mistakes ever to be made in wartime—yet, ironically,
an event even now celebrated for its heroism. This was the famous
Charge of the Light Brigade.

The charge occurred on October 25, 1854, during one of the major
battles of the war when British, French, and Turks squared off against
Russians over control of the main supply port being used by the British,
a city called Balaklava. To defend the port, the Turks had stationed a
small artillery force, backed up by a British force situated on high
ground.

The Russians eventually attacked Balaklava, sweeping past the
Turks and taking over some of the high ground. A further advance by
the Russians was stopped by a regiment of 500 British Highlanders,
under the command of Sir Colin Campbell. General Sir James Scarlett,
leading a regiment of heavy cavalry, drove the Russians off the high
ground.

Lord Raglan, commander of the British troops, watched the battles
from a distance. When he saw the retreating Russians trying to take
captured Turkish artillery with them (an affront to Turkish and British
honor), he called on the third element of the British forces—the brigade
of light cavalry commanded by Lord Cardigan.

Using a Captain Nolan, Lord Raglan sent an order to stop the
Russians from taking the Turkish cannons. But Nolan got the order
muddled, and what emerged was a directive to prevent Russia from
leaving with any cannons—Russian or Turkish.

The order first went to Lord Lucan, overall commander of the
cavalry, but he could not see the Turkish cannons; he saw only the
Russian cannons which were on high ground on the other end of
the valley. To get to these cannons, however, he would have to lead

*From "Crimean War" in *Encyclopaedia Britannica,* 14th edition (1929).

his men straight down the valley under the eyes and guns of the Russians.

Lord Lucan, realizing something was wrong with the order, objected, but he and Raglan were feuding, and when the command simply came down again, Lucan did not want to discuss it further. Besides, Lord Lucan intended to send the order on to Lord Cardigan to be carried out.

Now Lord Cardigan,* receiving the directive, also realized something was wrong. However, although he and Lord Lucan were brothers-in-law, they, too, were feuding. So when Cardigan questioned the order, Lucan simply repeated it as he had heard it from Raglan.

Cardigan, too proud to question the order further or to resist and be branded a coward, finally gave the command to his men to attack the Russian position across the valley. The Charge of the Light Brigade was on.

With Russian soldiers shooting down at the Light Brigade as it rode through the valley, the British casualties were fearful. Of the 673 men in the brigade, 113 were killed and 134 were wounded or captured. The casualties would have been even more devastating if a nearby French cavalry unit had not come to the rescue and attacked the Russians.

The battle for Balaklava eventually ended in a Russian victory, but it meant little. In six weeks the Russians left their positions, and later the allies claimed victory in the war when they captured the naval base at Sebastopol.

But the Charge of the Light Brigade had one unforeseen effect, an effect error sometimes has. People turned the error around to suit their needs—in this case an eagerness to find heroes in a wasteful war. The public treated what happened as a victory for courage rather than a defeat through stupidity and blunder. The surviving members of the Light Brigade and Lord Cardigan, never known as a likable man, became heroic figures when word reached Britain about the brigade's willingness to obey orders at all costs.

The Light Brigade was further glorified in a poem by the British poet, Alfred, Lord Tennyson, with that famous line "Theirs not to reason why, / Theirs but to do and die."

Thus, the Charge of the Light Brigade, a result of loused-up communication and fouled-up leadership, became a symbol of courage and

---

*Yes, the cardigan sweater was named after Lord Cardigan.

derring-do when in reality it should have entered history as a supreme example of derring-don't.

THE FIRST CIVIL WAR DEATH

The firing on Fort Sumter in South Carolina on April 12, 1861, is considered the opening battle of America's Civil War. But no one died in the thirty-four-hour skirmish, thirty-three hours of which involved consecutive cannon fire. That first fatality in the Civil War came only after the Union army at the fort had surrendered.

On April 14, while Union soldiers were firing a salute to the flag before evacuating the fort, an accidental explosion killed a Union soldier, Gunner Daniel Hough. He was a private in Battery E of the 1st U.S. Artillery and had survived the first battle of the Civil War only to die by mistake during the lull following the fighting.

THE *TITANIC:* SINKABLE? UNTHINKABLE!

Consider the shock, the disbelief of those who lived in that era.

The largest and most powerful ocean liner ever built until 1912 . . . designed with virtually every luxury then known . . . so oriented to the rich and powerful it had not one but two exclusive staterooms named and reserved for millionaires . . . and built—as unbelievable and as haughty as it sounded—to be unsinkable.

Yet on its maiden voyage this magnificent floating luxury hotel, designed and built and staffed and outfitted with the greatest of care, strikes a small iceberg and in two hours sinks with more than 1,500 passengers and millions of dollars lost.

And adding to people's disbelief was the fact that the *Titanic,* ballyhooed by its builders as unsinkable, didn't last three days at sea. How, everyone asked, could this happen?

The story of one of history's most embarrassing, tragic errors started in 1907, when British shipping magnates, uneasy watching the Germans capturing more of the transatlantic business with newer ocean liners than theirs, devised a plan to build a series of luxury ships that in size and scope would overwhelm the competition. In 1911 the first of those ships, the *Olympia,* the sister ship of the *Titanic,* was finished.

A year later the *Titanic* was ready. It was widely hailed as "the Wonder Ship" and "the jewel" of its builders, the White Star Line. Standing eleven stories tall, 883 feet long, and 92 feet wide and weighing 46,000 tons, it was made fractionally bigger than the *Olympia* so it would be the largest ship afloat. Its designers thought of one other attribute that would set the *Titanic* apart from all other ships: It was built to be unsinkable.

This was done by dividing the *Titanic* from prow to stern into sixteen watertight compartments. If water ever entered the compartments, an electrical switch that would close doors between them could be activated, thereby keeping the *Titanic* afloat.

There was only one problem with this theory—and here the error was made: The *Titanic* was designed to be unsinkable with any two of the compartments flooded—but not more than two. The designers did not figure on more than two compartments ever flooding. The thinking was similar to someone's driving a car with two spares to take care of all contingencies since two flats at any one time were rare. But what if something happened to the third or fourth tire? Yet, on as slight a thread of reasoning as that, the *Titanic* was touted as unsinkable.

And not only its builders made the claim. The technical journal *Shipbuilder,* in a special 1911 edition, noted the watertight compartments and the electronically controlled doors and said that with such a combination "the vessel [is made] practically unsinkable." Few, however, noticed the hedging in the use of the phrase *practically unsinkable.*

Another aspect of the *Titanic's* design was also overlooked in the claims about its unsinkability. The transverse bulkheads that divided the ship into its separate "watertight" compartments only reached the deckheads. This meant that water rising as high as the top of the bulkhead would pour over into the next compartment.

One final point: The most vulnerable part of a ship is its underside. The *Titanic* was no different. The ship's designers had reinforced the bow of the ship, but there was not much they could do with the sides of the ship below the waterline. They felt that as with most ships, problems, if they came, would be from frontal collisions, and for this the *Titanic* was prepared. But not for a gash along its underside.

Yet the unsinkable *Titanic* sank because of the erroneous planning in each of these areas: the number of watertight compartments, the height of the transverse bulkheads between them, and the vulnerable underside of the ship. Changes in any one of these three areas might have saved the ship called before its launching the Queen of the Ocean. But on April 14, 1912, before midnight, these very elements came together.

The *Titanic* was traveling at twenty-two miles per hour through ice-dotted waters in an attempt to break the record for a transatlantic crossing. Suddenly the fast-moving *Titanic* encountered a small iceberg (part of the iceberg may actually have been black, since when one breaks free of a glacier, the broken side is black). To avoid hitting it, the *Titanic* was swung to the left, but the iceberg caught the *Titanic* below the waterline, shearing a gash 300 feet in length 10 feet above the level of the keel.

Although, as mentioned, the *Titanic* could stay afloat with any two compartments flooded, the gash left six of the sixteen compartments filling with freezing water. Just forty-eight hours into its first voyage as the most publicized ship of its day, the *Titanic* was finished within the ten seconds it took for the iceberg to slash the ship's underside. A little more than two hours later—at 2:20 in the morning—the *Titanic*, tilted almost vertical, plunged nose down into 13,000 feet of water.

While ship designers, builders, and some of the major businessmen of the day parroted the belief that the *Titanic* was unsinkable, at least one of the passengers didn't—and the belief saved her life and that of her family. Eva Hart, who was seven at the time, was a survivor who later testified that her mother had had a premonition of disaster ever since the family had to transfer to the *Titanic* from another ship, the *Philadelphia,* because its crew was on strike. When the mother then saw a newspaper headline referring to the *Titanic* as unsinkable, she felt her premonition had been confirmed. "Now I know why I'm frightened. This is flying in the face of God," she told her family. Her fear about the *Titanic* kept Eva Hart's mother awake each night of the trip so that when the *Titanic* struck the iceberg at midnight, her mother heard the faint collision sounds and was able to wake the family and get them to the deck immediately.

Eva Hart's mother was certainly more ready for a collision than the ship's builders, who made provision for only sixteen lifeboats. While this was the regulation amount, the regulation itself was faulty: It called for ships displacing 15,000 tons or more to have a minimum of sixteen lifeboats, but in a classic Catch-22 of error, no one thought to increase the minimum number of lifeboats for the 46,000-ton *Titanic.*

The *Titanic* stands as a symbol of man's pride, his faulty attempts to control nature and dominate his environment, and his eventual defeat at the hands of his own error. It is therefore fitting that the *Titanic* was the first ship in distress from which the SOS signal was ever transmitted.

Finally, the story of the *Titanic* shows how one error begets another. Because of a garbled transmission, on April 15, 1912, a number of hours after the *Titanic* had sunk and now lay at the bottom of the ocean with 1,500 people drowned, the afternoon *New York Sun* appeared on the streets with a banner headline that proclaimed: ALL SAVED FROM TITANIC AFTER COLLISION.

TITANIC FOOTNOTES

The sinking of the *Titanic* is not the only time the biggest ship of its day was lost at sea. In the mid-nineteenth century the largest ship afloat was the 240-foot-long *President.* When launched in 1840, it was one of the first ships ever conceived, designed and built specifically to be a passenger liner. On just its third trip across the Atlantic, it left New York and was never heard from again.

In 1981 an expedition was sent to find the remains of the *Titanic* in the North Atlantic in 13,000 feet of water. Twenty-eight psychics announced that they knew where the *Titanic* would be found. Each psychic gave a different spot. The *Titanic* was finally located four years later.

Before the *Titanic* hit the iceberg, it was experiencing another problem which could also have led to its destruction. The *Titanic* was actually on fire. The U.S. Senate investigation of the sinking uncovered the fact that in one of the bunkers where coal was stored, dry coal at the bottom of the heap caught fire and continued to smolder in the face of attempts to put it out. One of the *Titanic*'s crew spent all his time trying to control the fire. The ship's captain knew about the problem and could have stopped the voyage to deal with it but chose not to. The persistent fire, in which flames raged beneath tons of coal, was still going strong when the *Titanic* hit the iceberg.

THE BLUNDER THAT COST GERMANY WORLD WAR I

By January 1917, after several years of intense fighting between Germany and the Allies, the Great War had become a bloody, costly stalemate. The key to an end to the war became the United States. A nearly bankrupt Britain desperately needed the United States to enter the war on its side; Germany wanted to keep America neutral as long as possible. President Woodrow Wilson, wishing to be the peacemaker, resisted all attempts to bring America's army into a conflict that had caused more than 1 million casualties.

But then, in April 1917, Wilson suddenly acceded to the wishes of Britain and France and sent America into the war. A year and a half later, Germany was defeated, forced to sign a peace treaty on November 11, 1918, with harsh terms imposed by the victorious Allies.

What had made Wilson change his mind? A number of scholars, including historian Barbara W. Tuchman, attribute Wilson's decision in large part to a telegram known as the Zimmermann telegram—one of the great blunders in history.

Arthur Zimmermann, a fifty-year-old bachelor with a bushy reddish brown mustache, was a nonaristocrat in an aristocratic German foreign service, the first person of middle-class origin to be appointed foreign minister. He had joined the Foreign Office in 1902, rose to undersecretary by 1911, and on November 22, 1916, found himself named foreign minister. The American press greeted his appointment, the *New York Evening Post* even headlining an article on him OUR FRIEND ZIMMERMANN.

Zimmermann had a strong interest in America and even considered he knew and understood the American character on the basis of one strange factor: He had once taken a train trip across the American continent.

For all that, Zimmermann's one thought on taking over as foreign minister was to get Mexico, which was not in the war, and Japan, which was then in the war as an ally of Britain, to form an alliance with Germany. The German military had to launch an intensive U-boat campaign to bring about Britain's final defeat, and Zimmermann feared U.S. entry into the war unless he could keep America neutral. But if this failed and the United States entered the war against Germany, he wanted to be able to call on Mexico and Japan to fight against the Americans to keep the United States out of Europe. Zimmermann's incentive for Mexico to attack the United States would be Germany's pledge to help it recapture territories it had lost to the United States. He felt that Mexico would then help draw Japan into the war alongside Germany, something Germany had been trying to do for several years.

The question before Foreign Minister Zimmermann was how to

conduct such delicate negotiations with faraway Mexico and still keep discussions secret.

At first Zimmermann intended to send instructions by letter to the German imperial minister in Mexico for him to propose to the president of Mexico an alliance with Mexico and Japan and outlining Germany's intention, if necessary, to support Mexico in an attack on the United States. Zimmermann intended to send the letter by a German submarine which was scheduled to depart on January 15, but shortly before departure the trip was canceled. He then decided to send the instructions by telegram. But by what method?

It was then that Zimmermann made his fateful blunder.

The method he hit on was to use the communications channel President Wilson had made available to Germany for the purpose of keeping in touch during peace negotiations. The cable route began in the American Embassy in Berlin, passed through Britain, and wound through the State Department in the United States. This was the most direct way then available for Germany to reach its ambassador in Washington, who could then relay any message to Mexico. Although Zimmermann's message would violate the purpose of the channel, he felt that since the telegram would be sent in a German-devised code for the German ambassador in Washington, he had little to fear in its contents' being discovered.

Thus, Zimmermann was confident enough in the German code that he was willing to use the American cable route to begin negotiations to carry out a military alliance against America. What Zimmermann did not know, however, was that British naval intelligence had broken the German code. In fact, as Zimmermann's telegram came across the American cable in Britain in code, the people in British naval intelligence had little problem in picking up and, after some work, deciphering probably one of the most damaging telegrams ever sent in wartime. Here is the full text of the telegram sent by Germany's foreign minister to the German imperial minister in Mexico on January 16, 1917:

WE INTEND TO BEGIN UNRESTRICTED SUBMARINE
WARFARE ON THE FIRST OF FEBRUARY. WE SHALL
ENDEAVOR IN SPITE OF THIS TO KEEP THE UNITED
STATES NEUTRAL. IN THE EVENT OF THIS NOT
SUCCEEDING, WE MAKE MEXICO A PROPOSAL OF

ALLIANCE ON THE FOLLOWING BASIS: MAKE WAR
TOGETHER, GENEROUS FINANCIAL SUPPORT, AND AN
UNDERSTANDING ON OUR PART THAT MEXICO IS TO
RECONQUER THE LOST TERRITORY IN TEXAS, NEW
MEXICO, AND ARIZONA. THE SETTLEMENT IN DETAIL IS
LEFT TO YOU.

YOU WILL INFORM THE PRESIDENT [OF MEXICO] OF THE
ABOVE MOST SECRETLY AS SOON AS OUTBREAK OF WAR
WITH THE UNITED STATES IS CERTAIN AND ADD THE
SUGGESTION THAT HE SHOULD, ON HIS OWN INITIATIVE,
INVITE JAPAN TO IMMEDIATE ADHERENCE AND AT THE
SAME TIME MEDIATE BETWEEN JAPAN AND OURSELVES.

PLEASE CALL THE PRESIDENT'S ATTENTION TO THE
FACT THAT THE UNRESTRICTED EMPLOYMENT OF OUR
SUBMARINES NOW OFFERS THE PROSPECT OF
COMPELLING ENGLAND TO MAKE PEACE WITHIN A FEW
MONTHS. ACKNOWLEDGE RECEIPT.

                                              ZIMMERMANN

As Pulitzer Prize winning historian Barbara Tuchman notes in her book on this aspect of World War I, *The Zimmermann Telegram* (Macmillan, 1966), "Zimmermann's choice of telegraph route could hardly have been more inappropriate, under the circumstances, yet it was perhaps predestined by the German character. The fatal German assumption of superiority, superior right, superior cleverness, led him straight into it."

At first British intelligence delayed bringing the telegram to U.S. attention out of fear it might tip off the Germans that their code had been broken. When the British finally brought the telegram to Wilson's attention on February 24, he was greatly indignant. The telegram was finally made public on March 1 to dramatic headlines in the American press and to angry debate in Congress. Yet the Germans still had the opportunity to defuse the situation by denying the telegram, and since Great Britain still wanted to keep secret its ability to break the German code, the U.S. government would have been hard pressed to show up the German denial.

It was then that Zimmermann committed a second blunder. As Tuchman writes, "Unbelievably . . . to the 'profound amazement and relief' . . . of everyone concerned, Zimmermann inexplicably admitted his authorship."

The admission astounded America, shattering "the indifference with which three-quarters of the United States had regarded the war until that moment." As one American newspaper editorialized, "The issue shifts from Germany against Great Britain to Germany against the United States." With Germany plotting to attack U.S. territory, there could now be no longer a question of U.S. neutrality. Even Wilson was swept along in the outrage against Germany. On April 2, 1917, he told a joint session of Congress to "declare the recent course of the Imperial German Government to be in fact nothing less than war against the government and people of the United States" and urged that America "formally accept the status of belligerent." He specifically cited the Zimmermann telegram. Germany, he said, "means to stir up enemies against us at our very doors, the intercepted note to the German Minister in Mexico is eloquent evidence."

Wilson received overwhelming support. On April 6, 1917, the United States declared war on Germany and entered the First World War.

Tuchman considers the Zimmermann telegram not necessarily the deciding factor influencing Wilson to bring America to Britain's side, but it was "the last drop that emptied his cup of neutrality." A close confidant of Wilson's, Ray Stannard Baker, whom Wilson himself asked to write his official biography, went further. He stated that "no single more devastating blow was delivered against Wilson's resistance to entering the war."

The telegram had even more dramatic effect on American public opinion, which had been supportive of Wilson's desire to stay out of the war but now shifted drastically. A *Literary Digest* roundup of U.S. press reaction tells the story in its title: "How Zimmermann United the United States."

Summing up the significance of Zimmermann's blunder, Tuchman writes that had the telegram not been intercepted, something else the Germans did might have brought the Americans into the war, but time was of the essence, and too much of a delay might have forced the Allies to negotiate with Germany. "To that extent the Zimmermann telegram altered the course of history."

Ironically, even while the furor over the telegram was going on in the United States, Zimmermann was trying to find out how the message had been discovered—but to find out, he kept sending telegrams in the same code over the same cable to the same German diplomat in Mexico. Never once did he think that the code had been broken.

Zimmermann then still tried to bring Mexico into the war, assuring its president of huge sums of money with which to buy arms. But after some hesitation, the president of Mexico decided to remain neutral; according to the German minister in Mexico, the "premature publication" of the telegram had wrecked possibilities for such an alliance at the time.

With the possibility of a Mexican alliance gone, Japan far from willing to enter the fray on Germany's side and the United States declaring war on Germany, Arthur Zimmermann's days in high office were numbered. Within four months he lost his position when the chancellor of Germany lost his. After less than a year Zimmermann's career as foreign minister—and sender of telegrams in code—was over. In his brief term in office, however, Arthur Zimmermann had certainly left his mark on world events.

HOW A MALE SECRETARY BOTCHED THE WORLD WAR I ARMISTICE SIGNING

The armistice signed to end World War I may not have been a valid document. The male secretary who typed it made a mistake which rendered much of it illegible, but none of the signers noticed.

The text of the document to end the war that was said to end all wars was dictated by Marshal Ferdinand Foch to Henri Deledicq, a military clerk from French headquarters who was assigned to the railway car near Rethondes in the Compiègne Forest where the final armistice session took place early on the morning of November 11, 1918.

Shortly after 5:00 A.M., with the details settled, it was agreed to type the last page of the text immediately so that the signatures could be applied and hostilities ended as soon as possible. As Deledicq began typing the agreement, the sleepy clerk accidentally put some of the carbon papers in backward, and whole parts of the document emerged unreadable. Much of it was back to front. "I was too tired to notice," he recalled years later. As soon as he had finished typing, the German delegation signed the armistice agreement without reading it. Not one of the signers questioned what he was signing. Only some time afterward did an incredulous Marshal Foch find out about the error.

Years later the clerk was asked to describe Foch's reaction upon discovering the garbled document. Said our Henri: "He could hardly get over it."*

" 'FRIENDLY' SWORDCUTS": THE ACCIDENTS OF WAR

Accident and error play a key role in one of the strongest forces shaping history—warfare. John Keegan, a senior lecturer in war studies at the Royal Military Academy in Britain, details in his book *The Face of Battle* (Vintage Books, 1977) the many ways wars are affected—and lives are lost—not by fighting between armies but by accidents within an army.

"Accidental wounding is one of the major hazards of battle, and the desire to avoid it one of the principal reasons underlying the professional soldiers' much derided obsession with drill," he writes, noting that with close-packed groups of men equipped with firearms, "one's neighbor's weapon offers one a much more immediate threat to life than any wielded by an enemy." Indeed, at a battle such as Waterloo, Keegan reports that there exists "numerous authentic accounts of losses by 'friendly' fire—or even 'friendly' swordcuts."

Accidents, which have always caused a proportion of battle deaths and wounds, have increased with modern warfare. Contributing to this are firearms and artillery which explode on their users or are misdirected, more volatile and powerful explosives, tanks which because of hampered visibility have proved dangerous to the infantry accompanying them, and minelaying and mine lifting. Keegan cites the case of a British regiment which actually suffered more casualties in training accidents during World War II than at the hands of the enemy.

Furthermore, Keegan notes that with the increased mechanization of armies, young soldiers are more involved in driving tanks, jeeps, and other vehicles along hazardous roads, with the result often of an alarming rate of fatal mishaps. Says Keegan: "During quiet weeks in the

---

*Marshal Foch engaged in some foul-ups of his own. His overreliance early in World War I on a military strategy of attack rather than defense almost destroyed his career. He badly misjudged German firepower and as a result participated in French offensives that failed in 1914, 1915, and 1916. He was finally relieved of his command at the end of 1916. But as the *World Book Encyclopedia* notes, "One of Foch's great strengths . . . lay in his ability to learn from his earlier failures." By 1918, he had worked his way back into favor, and was promoted to marshal three months before the armistice. (Foch was eerily correct about one thing. In 1919 he predicted another war within twenty years.)

THE REIGN FALLS EVERYWHERE

Error is no respecter of rank; it reigns over every area of our lives. Consider the case of a president of the United States and his misuse of the English language.

Warren Gamaliel Harding (1865–1923), the twenty-ninth president, had been the editor of a newspaper in a small Ohio town. Because of this background, he believed he had a special ability to write and therefore liked to compose his own speeches. This was an admirable thing to do, except that Harding invariably blundered in his choice of words. H. L. Mencken termed Harding's speeches "the worst English I have ever encountered."

At one point, a furious Mencken was moved to comment about Harding's misuse—and abuse—of the English language: "It reminds me of a string of wet sponges; it reminds me of tattered washing on the line; it reminds me of stale bean soup, of college yells, of dogs barking idiotically through endless nights. It is so bad that a sort of grandeur creeps into it."

What had enraged Mencken? Here is a sampling of Hardingisms, as gathered by Paul Boller, Jr., in *Presidential Anecdotes* (Oxford, 1981):

> *"I like to go out into the country and bloviate."
> *"We must prosper America first."
> *"I would like government to do all it can to mitigate."
> *"Progression is not proclamation nor palaver. It is not pretense nor play on prejudice. It is not personal pronouns, nor perennial pronouncement. It is not the perturbation of a people passion-wrought, nor a promise proposed."
> *"America's present need is not heroics, but healing, not nostrums but normalcy, not revolution but restoration, not agitation but adjustment, not surgery but serenity, not the dramatic but the dispassionate, not experiment but equipoise, not submergence in internationality but sustainment in triumphant nationality."

When Harding died, poet e. e. cummings wrote, "The only person who ever committed six errors in one sentence has passed away."

Vietnam campaign, traffic accidents often killed more American soldiers than did the Viet Cong."

It has been said from old times that a battle is a succession of mistakes and that the party which blunders less emerges victorious.

—Lieutenant Commander Masatake Okumiya
of the Japanese Navy,
quoted by Gordon Prange, *Miracle at Midway*

KRISTALLNACHT

On November 9, 1938, the Nazis launched a massive countrywide pogrom that not only led to the arrest of 30,000 German Jews but caused such devastation to synagogues (191 set on fire) and to Jewish shops, homes, and buildings (more than 1,000 torched or destroyed) that it was called *Kristallnacht* ("Crystal Night" or "Night of the Broken Glass").

The Nazis had been looking for a pretext to launch an all-out attack on the Jewish community of Germany. They found it in the shooting of a German embassy official in France two days before. On November 7, 1938, in Paris, a young German Jew, distraught over learning his parents had been driven destitute out of Germany and into Poland, went to the German Embassy and shot a German official. Herschel Grynszpan had wanted to kill the German ambassador but fired at the first German he encountered. That person turned out to be Ernst vom Rath, a minor German Foreign Office official, who died two days later, giving Hitler the opportunity for which he had been looking. Unbeknownst to Grynszpan, however, Rath was at that very time under investigation by the Gestapo for his anti-Nazi attitude and was himself opposed to anti-Semitism.

THE MAGINOT LINE: THE SHORTEST DISTANCE BETWEEN TWO GERMAN POINTS

One of the axioms of historians is that military planners plan for the last war. The truism of this fault in the military mind is nowhere better seen than in the story of the Maginot Line.

The Maginot Line was a fortified line of defense built by the French along their eastern border with Germany. Constructed after World War I, which had seen grievous trench warfare, the Maginot Line was in essence a trench to end all trenches. Aboveground the French military built forts, pillboxes, barbed-wire entanglements, tank barricades, and machine-gun nests. Underground they constructed a multilevel air-conditioned fortress, complete with hangars, hospitals, ammunition storerooms, living quarters for officers and soldiers, a telephone system, even a gymnasium, movie theater, and subway. The whole system was tied into a series of little Maginot lines, lesser forts built apart from but connected underground to the main Maginot Line.

This "supertrench," which took years and millions of dollars to build, was considered by the French military to provide France with an impregnable defense against a German attack. But in June 1940, when the Germans did attack, they simply invaded France by driving through Belgium and pouring into France from the north, thereby skirting around the Maginot Line, which they captured from behind. It took Germany just three weeks to wind up with France, its Maginot Line—and the Maginot Line movie theater.

In the 1950s France overhauled the Maginot Line to prepare it for another use—refuge in case of atomic attack.

THE BLUNDERS AT PEARL HARBOR

December 7, 1941, lives not only in infamy but in the annals of error. For the disaster that Japan was able to inflict on the United States that day has been attributed in large part to blunders by the U.S. navy high command.

—Although a surprise to the United States, the attack was no surprise to U.S. naval intelligence. The navy had broken the Japanese code and knew that Japan planned an attack but did not know where. Then, on Saturday, December 6, Mrs. Dorothy Edgers, an employee who had been with the Office of Naval Intelligence for just a month, decoded an intercepted Japanese message that implied Honolulu was to be the target. She was told by her superior officer that the message needed more work and could wait until Monday.

—Virtually the entire U.S. fleet based in Pearl Harbor was in the harbor on December 7, even though the usual practice was to deploy many of the vessels on patrol so that the fleet would not be congregated in any one place at any one time. But a practice that had grown during peacetime was to allow shore leave for as many men as possible over the weekend. In Hawaii this became an especially strong tradition because many of the officers and sailors had brought their families to live in the nearby Honolulu area. But the widespread granting of shore leave ensured that most of the fleet would be in Pearl Harbor on a weekend—a dangerous mistake and one on which the Japanese fully capitalized by attacking on a Sunday.

But let us not overlook the substantial blunder that Japan made at Pearl Harbor: Although the Japanese destroyed much of the fleet anchored in port that day, they did not attack the U.S. ships at sea, nor did they follow up their initial assault on Pearl Harbor in subsequent days, which would have dealt a knockout blow to U.S. naval operations. Not quite believing their good luck and too worried about a trap, they allowed their December 7 attack to serve as their only bombardment of Pearl Harbor. By failing to come back to wipe out the rest of the U.S. naval arsenal, they permitted the U.S. forces to regroup. Many of the ships spared at Pearl Harbor were used later at the Battle of Midway, where the U.S. navy defeated the Japanese navy to turn the tide in the war in the Pacific. In this very important respect, then, Japan was not the victor but the loser of Pearl Harbor. Japan's mistake was to launch an attack that did not prevent America from winning a war. Pearl Harbor only ensured that America would enter the war.

HITLER'S MISS

While Adolf Hitler was able at one point to capture control of much of Europe, the *Führer* committed at least one glaring mistake that may have cost him total victory.

At the end of World War II the U.S. military was astonished to discover that Germany had no atomic weapons and was not far along in atomic research. This glaring lapse has been attributed to Hitler's anti-Semitism. His hatred of Jews had caused many important Jewish physicists who could have been used to develop Germany's atomic weaponry to flee Nazi control. Several of them—such as Albert Einstein, Lise Meitner, and Edward Teller—were instrumental in developing America's atomic bomb. In fact, Hitler disliked the entire field of physics, labeling it "Jewish physics." Whenever the subject came up, the *Führer* turned furious.

WHEN SAINTS TURNED OUT TO BE SINNERS

In December 1941, with the German army only fifteen miles out of Moscow, Joseph Stalin exhorted Soviet soldiers about to go off to battle to do their best to defend Mother Russia. In his Red Square address, Stalin stirred his troops by invoking the memories not of Marx and Engels but of Russia's old saints. It was a reminder that even in atheistic Communist Russia there is much devotion to the saints venerated by the Russian church. Indeed, still in existence among Russians is a deeply held belief that because of the protective influence of Russia's saints, no other country can conquer the Soviet nation.

It was then all the more shattering for many Russians when, in 1981, Soviet forensic experts established conclusively that relics of three of Russia's most revered saints were really the bones of Mongol invaders, their very adversaries.

As related in the *Washington Post* of August 7, 1981, the skeletons, which had been kept in ancient Russian Orthodox religious centers near Moscow, had long been thought to be those of Bishop John, St. Euphemia and St. Euphrosinia—three fourteenth-century Russian saints said to have miraculously shielded their communities from Tartar hordes. In examining the relics, however, Russian scientists discovered evidence that revealed a totally different story. The skeleton thought to be that of Euphrosinia had a skull with the flat features characteristic of the Mongols and a body that had only been four feet eight inches. The Bishop John relic also had the features of a Mongol, and its bones were those of a young man no older than thirty-five (Bishop John was said to have lived to a ripe old age).

As for Euphemia, who died in 1404 at the age of eighty-eight, the relic was actually composed of three skeletons, one of which was a child; the other parts of the skeleton had too many ribs and bones to be that of a single body.

The irony in this error of religious devotion is that in Russian eyes the bones that had been so venerated for so long were those not of saints but of sinners.

SUMMING UP SUMMITRY

The top leaders of the United States and the Soviet Union have gotten together periodically since World War II, in what have been called summit meetings, to try to solve amicably some of the most pressing issues facing the two world powers. These high-level sessions have usually been accompanied by high hopes, followed by low results. Once, during President Dwight David Eisenhower's meeting with Soviet Premier Nikita Khrushchev at Camp David, a reporter covering the talks inadvertently summed up the feeling of those observing these sessions by writing: "A private meeting was held but no details were given as newsmen were bored from the meeting."

THE BAY OF PIGS

The Bay of Pigs may well be one of the major mistakes of American foreign policy. This was the episode during the Kennedy presidency in which the United States tried by covert support of Cuban dissidents to overthrow Fidel Castro and rid Cuba of Communist control. What marked the brief attack, launched in Cuba's Bay of Pigs, was the expectation by both the Cuban insurgents and their American supporters that this would be a brief and easy battle in which the people of Cuba would rise up and help defeat Castro. The quick victory by Castro's forces stunned everyone (among the reasons the Cuban exiles lost were that they were poorly trained and were given out-of-date maps).

William Manchester, a Kennedy biographer, called the episode "the gravest blunder of his career." JFK later assumed the responsibility for the blunders in planning and carrying out the attack, noting wryly one truth about error: "Victory has many fathers, but defeat is an orphan."

But Kennedy was not the only one responsible. Eisenhower had started the escapade and urged it on Kennedy, the CIA planned it, and the Joint Chiefs of Staff unanimously predicted success. One result of the debacle, according to Manchester, was that Kennedy "lost all illusions about the Joint Chiefs' infallibility."

Two weeks after the Bay of Pigs, Kennedy's standing in public opinion polls actually went up—to an unprecedented 83 percent favorable rating.

THE KEY TO WATERGATE WAS THE KEY

The attempted burglary of the Democratic National Headquarters in the Watergate complex in Washington, D.C., on June 17, 1972, eventually led to the resignation of Richard Nixon as president of the United States—the first time in the history of America that a president had resigned. Although much has been written about the less than successful burglary and the subsequent attempt to cover up its connection to the Nixon administration, William Goldman, a screenwriter employed to write the movie about the definitive book on the story *(All the President's Men),* uncovered a little-known aspect to how it all happened.

Writing in his own book *Adventures in the Screen Trade* (Warner Books, 1983), Goldman tells how his preliminary research of the burglary revealed "the inept quality of so much of what went on."

He found, for instance, that the break-in of June 17 was not the first try by the group to burglarize Democratic headquarters in the Watergate. It had attempted to do so several times previously but "kept goofing it up."

"Once, they got trapped and had to hide for the night in an empty room in the complex," Goldman relates. "Better than that, another attempt failed because the keys they had made to get them into the Democratic National Headquarters didn't fit."

When the burglars realized the keys were not working properly, they decided to go back to where they had had them made. The problem was that they had had them made in Miami. But this fact did not seem to faze them. According to Goldman, "after their bungle some of them *went back* to Miami to have keys made all over again."

THE VIETNAM WAR IN ONE SENTENCE

Volumes have been written about the political and military blunders made in connection with the Vietnam War. The United States spent more than 50,000 lives and billions of dollars in a ten-year struggle that many people in America and among its allies questioned throughout the entire period. But there is one sentence, inadvertently said, that may sum up all the feelings of frustration held about the wisdom of leaders who advocated the war. The statement was uttered by the late Hubert Humphrey when he was vice president in Lyndon Johnson's administration. Said Humphrey: "No sane person in the country likes the war in Vietnam, and neither does President Johnson."

THE FUTURE OF THE WORLD—DOES IT REST ON PREVENTION OF AN ERROR?

What will the future history of our world be like, especially a world filled with nuclear weapons? According to the author of a best-selling book on the nuclear issue, the answer may hang on how well humanity can prevent an error from tripping all-out nuclear war.

In *The Fate of the Earth* (Knopf, 1982), a book which caused a great stir when issued, author Jonathan Schell points out that one of the ways in which a nuclear war could occur would involve "a wholly accidental attack, triggered by human error or mechanical failure."

Schell does not suggest such a scenario lightly. He cites three incidents during several previous years when United States nuclear forces were placed on alert and poised to respond to a nuclear attack because of error or foul-up. Twice this happened "because of the malfunctioning of a computer chip in the North American Air Defense Command's warning system."

Another time it happened "when a test tape depicting a missile attack was inadvertently inserted in the system."

In fact, because the United States and the Soviet Union are poised in anticipation of a planned nuclear attack, a mistake by human or computer may be what overrides the restraints now preventing nuclear war. Schell writes: "The greatest danger in computer-generated misinformation and other mechanical errors may be that one error might start a chain reaction of escalating responses between command centers, leading, eventually, to an attack."

Schell shows how one nation's computer, reading the mistake of another nation's computer, might trigger a state of emergency in one country which the other country would read as evidence that its own computer had not been malfunctioning but had been reading the situation correctly. The result could be a quickly escalating response on both sides, building up to an inescapable result—all-out nuclear war.

Thus, our world, which many scientists say began with an explosion (the big bang theory), would end in another colossal explosion—this one very possibly the result of error.

THE "WE BUILD 'EM TO LAST" DEPARTMENT:

THE LEANING TOWER OF PISA

Humanity's efforts to erect sturdy, secure structures ranging from one-story edifices to buildings soaring thousands of feet have encountered many problems and numerous failures. Bridges have collapsed. Buildings have fallen even before finished. Floors have given way. Roofs have caved in.

For the past six centuries a major symbol of faulty construction has been the famed Leaning Tower of Pisa. Almost from the beginning of construction in 1174, this bell tower in the small Italian town of Pisa, 150 miles north of Rome, has been moving toward collapse and resisting all attempts to stop the listing. Its architect had made the mistake of calling for only a ten-foot foundation, and when the ground shifted during construction, the building began tilting. Since completion in 1350, the tower has shifted seventeen feet from the perpendicular. Scientists have found not only that they are helpless to stop the leaning, which now continues at an average rate of one-quarter inch a year, but that they cannot even predict when the tower will finally topple. During 1982 the Italian government spent $10.5 million to stop further tilting, but still the building leaned another millimeter. The plans now call for trying to stabilize the tower by increasing the pressure of ground water around it. However, in 1934 a similar project that pumped concrete under the base actually sped up the tilting.

The centuries-old blunder in building the Tower of Pisa on such a faulty foundation has had one beneficial effect: A Nonleaning Tower of Pisa would probably have been torn down long ago to make way for a pizza parlor parking lot.

Error can be beautiful . . . and a tourist attraction.

CHAPTER II

Slip: Error in Medicine

The magazine cartoon shows a scene in heaven. One angel—with a halo and wearing a robe—is saying to another angel, "The last thing I remember the doctor saying was 'Oops.'"

—Discovered by the author
taped on the wall of
a hospital emergency room

The medical profession, dealing as it does with matters of life and death, is invested in our minds with an awe and a hope. The awe is the reverence we feel for the doctor's knowledge and training; the hope is the yearning we feel for his curative powers. For this reverence, we have historically held the doctor and the medical profession in high esteem; in return, the doctor has often held his fees equally high.

The general tendency has therefore been to consider physicians free of error and to view any mistakes in medical treatment as rare and especially dismaying. The truth is that slipups can be found in doctors' offices and hospital rooms as much as anywhere else in the world of human existence.

The history of medicine shows that not until relatively recent times have real progress been made in medical treatment and the medical profession been freed from much ignorance and error. There have been long and numerous periods when no new progress was made in medicine and the old errors were simply repeated. Dr. Howard Haggard, professor of physiology at Yale University earlier this century, wrote in a book on medical history that "the usual and hence normal state of medicine has been one of almost complete stagnation and complacency. Century after century has passed in which not one single new fact was gained in the entire field of medicine—indeed, when the search for the new was rigidly forbidden."

According to Dr. Haggard, the major error in medicine in the past has been the error that goes beyond mere ignorance: "It is the error of complacency . . . to cease the search for knowledge and to crystallize both fact and fallacy into dogma."

The medical world finally made its greatest strides during the past 150 years, with advances in surgery, anesthesia, antibiotics, antiseptic procedures, and diagnostic equipment. Only then was the profession uplifted and the image of the physician improved. (One need only realize that until the advent of anesthesia many surgeons were also barbers.) But even with such progress, mistakes continue to be a part of the medical scene—to the surprise of doctors and the discomfort, if not deaths, of patients.

A revealing view of physician fallibility can be found expressed in a recent book by a doctor, no less—Dr. David Reuben, a psychiatrist and best-selling author. Advising his readers to get second opinions in case of serious illness, he declares in *Dr. David Reuben's Mental First-Aid Manual* (Macmillan, 1982), "Don't trust any single doctor." He recounts how he "will never forget the colleague" who treated the wife of a minister for syphilis—which she did not have. "My colleague won't forget either because when she took him to court the jury decided that he had to pay her $250,000 for that careless diagnosis."

Dr. Reuben then touches on a point people overlook or hide from: the need for the patient to be as aware of and alert to possible error in medicine as in any other phase of life.

"Doctors make mistakes—no one can deny that," Dr. Reuben writes. "And *you* don't want to be one of those mistakes."

Medicine in Earlier Times

In earlier times medicine was ruled more by emotion and whim than by careful testing and close observation. The ancient—and not so ancient—art of medicine was truly more art than science. As a result, the world of premodern medicine was a breeding ground for errors as well as for germs.

Indeed, one definition of error—in fact, in many dictionaries it is listed as the first definition—is a mistaken belief in what is untrue. The history of medicine is replete with medical beliefs about diseases and their treatment that were later shown to be not only false but often harmful. Most times the original error was due to lack of enough information and experience. But sometimes it was due to the human being's tendency to prove anew the wisdom of Murphy's Law: Given two paths to follow or two decisions to make, people will invariably go the wrong way.

In this chapter and in the next on science we will see how, especially in previous ages, the human being has held mistaken beliefs about some very critical medical and scientific subjects—and how those flawed ideas adversely affected humanity for centuries.

THE SEARCH FOR A CURE FOR MALARIA

Malaria, which is carried by mosquitoes, is estimated to have killed more people than has any other disease. Beginning in the tropics, it has swept the world and affected history (among its many victims: Alexander the Great at the height of his career).

In the centuries-long search for a cure, numerous mistakes were made along the way. The result was even more deaths than necessary. Some of the errors:

> —Galen, a noted and influential Greek physician of the second century, believed the high fevers caused by malaria were due to a disorder in the four humors of the body (the humors were yellow bile, black bile, phlegm, and blood). He wrote that the only way for the physician to restore the patient's normal balance of humors was by bleeding (which would rid the body of "corrupt humors") or by purging (which would help empty the body of poisons)—or better yet by doing both. Although bleeding and purging served only to weaken a weakened body further and to hasten death, Galen's remedy was widely accepted for the next 1,500 years. Ironically, since Galen's ideas were followed mainly by the wealthy, who had doctors, poor people

tended to fare somewhat better. As one history of
medicine notes, "Lucky indeed were the country folk
and townspeople too poor to afford the ministrations of
the medical profession. These common citizens turned
to witchcraft, a form of treatment often less harmful
than the orthodox methods."
—Among other remedies suggested as cures for malaria:
In the thirteenth century a churchman advocated that
the patient's urine be mixed with flour, made into
dough, and formed into seventy-seven small cakes,
which were then to be taken to an anthill before
sunrise and thrown in. It was believed that as soon as
the insects ate the cakes, the fever would vanish. . . . A
treatment suggested in England was to rub the chips
from the gallows on which a criminal had been
executed on a malaria victim. . . . A Spanish-originating
cure involved the patient's drinking a glass of brandy
spiked with three drops of blood from the ear of a cat
and a touch of pepper.
—When a cure was stumbled across, not everyone used
it, even when forgoing the cure meant death. The
remedy that did work had an unlikely source—the bark
of the cinchona* tree. Unbeknownst to those who
began using it in the seventeenth century, this bark
contained quinine, an effective drug against malaria.

One of those who erred in forgoing the cinchona
was Oliver Cromwell, who died of malaria in 1658. The
Puritan leader refused to take the bark because he
thought its use part of a Catholic plot. Since the Jesuits
had been the first Europeans to promote cinchona for
malaria, many Protestants viewed the bark as a
Catholic campaign to poison non-Catholics. Cromwell
made the fatal mistake of confusing medicine with
martyrdom. (One leader who did take the bark and

*The remedy for malaria is misspelled. Carolus Linnaeus, a Swedish botanist, named
the tree producing the bark cinchona after the countess of Chinchón in Spain, who was
said to have been cured by the use of the bark and who then gave it to the peasants
living on her estate. But the chemist followed a misspelling of the Spanish name when
it was transposed into Italian as "Cinchón" instead of "Chinchón." Only after his death
was the mistake discovered, but it was too late to correct it.

thereby survived an attack of malaria was George
Washington.)
—And a scientist's mistake enabled malaria to continue
its scourge of humanity. Armand Seguin, a French
chemist who in 1804 had devised a method for
extracting morphine from opium, turned his attention
to the cinchona bark and came to the faulty conclusion
that it was successful against malaria because it
contained gelatin. Despite inadequate data, he
published his erroneous findings. "For years thereafter,
many physicians reading Seguin's paper adopted
clarified glue to treat their malaria patients," writes Dr.
Steven Lehrer in *Explorers of the Body* (Doubleday,
1979). "This enormous error is all the more surprising
because Seguin was actually quite a facile chemist."*

WAYS OUR PRESENT LANGUAGE SHOWS
MEDICAL ERRORS OF THE PAST

—Malaria is caused by diseased mosquitoes, but the name
comes from *mala aria,* "bad air," because it was
thought that evil spirits, which traveled at night,
brought disease into the body through the inhaling of
night air.
—Genes carry the heredity of a person, but it used to be
believed that the blood fulfilled this purpose, a fact we
can see in the current usage of *bloodline, bad blood,*
and *blood relations.*
—While the brain is the source of thought and therefore
of desire, the heart was once seen as the source of
sexual passion. This can be seen in the use of
sweetheart, broken heart, and *heartbeat* in relation to
love.
—The liver is so called because at one time it, rather
than the heart, was considered the sole blood-giving
organ that sustains life.

*Seguin made another career error. He did not publish his method for extracting
morphine from opium for ten years after his discovery—and by then a German chemist,
Friedrich Sertürner, had published his technique (in 1806). Because of this lapse in both
judgment and time, it is the German, not Seguin, who is considered the discoverer.

DOCTORS ON SEX

The sexually transmitted disease of syphilis, which, if unchecked, can lead to blindness, insanity, and death, became one of the scourges of the Western world when it was first encountered in the late 1500s. But the doctors of the day made the disease even worse because of their own erroneous theories and treatments.

For example, the most widely accepted medical treatment for syphilis was mercury, which had to be taken over an agonizingly long period of time. In fact, treatment was so prolonged it led to an expression: "One night with Venus/A lifetime with Mercury." Today, however, it is understood that mercury is ineffective against syphilis.

And then a leading British anatomist and surgeon of the eighteenth century advocated a theory which actually slowed progress in fighting the disease. John Hunter (1728–1793) believed that syphilis and gonorrhea were due to a single poison. Although both are transmitted sexually, they are different diseases. Hunter's experiments with the two and his theories about combating them together retarded understanding of both syphilis and gonorrhea.

Not until the twentieth century and the work of Dr. Paul Ehrlich were there a true understanding of and treatment for these afflictions.

THIS IS THE FATHER OF MEDICINE SPEAKING . . .

The rule is: be helpful to the patient; at any rate, do him no harm.

—Hippocrates, *Epidemics*

THE "POWDER OF SYMPATHY"

While many outrageous treatments have been tried without success, one remedy proved effective—but for the wrong reasons.

During the time of King James I of England (he reigned 1603–25), the medical fraternity employed what were called weapon-ointments and powder of sympathy to help heal battlefield injuries. A soldier's wound was cleaned and covered with a bandage, and a salve was applied *not* to the wound but to the weapon that had caused it. If the weapon could not be found, a piece of bloody clothing was dipped in a powder.

The bizarre treatments seemed to work. In fact, it was found that the wounds healed more quickly when the ointment was applied to the weapon or the powder to the clothing than when it was placed on the wound.

The reason for the success of this approach became clear only when it was discovered much later that the procedure kept the dirty, bacteria-infested ointments and salves of those days away from the wound, thereby preventing infection. When doctors previously treated the wound directly, they had unwittingly been causing the infections they tried to prevent.

THE HISTORIC BOOK WRITTEN ON BLOOD—AND CHEAP PAPER

William Harvey (1578–1657), an English physician and anatomist, was the first person to discover how blood circulates in the body—perhaps the single most important discovery in the science of physiology. Harvey published the results of his studies in 1628 in a little book entitled *De Motu Cordis*—or *An Anatomical Treatise on the Movement of the Heart and Blood in Animals.*

But from the way the book was published, it is apparent that Harvey did not realize the importance of his discovery. He sent the manuscript to an obscure German printer in Frankfurt am Main. The materials used in the publication were cheap, the paper was thin, and the book began to deteriorate quickly. The finished work itself was filled with so many typographical errors it appears that neither the printer nor Harvey bothered to proofread the copy before publication.

Yet this typo-filled book changed the course of all medical thought. Because our understanding of the causes of diseases and their spread depends on understanding the circulation of the blood, Harvey's discovery has been put by historians of medicine on a par with Newton's discovery of gravity. Except that Newton retained a better printer for his monumental *Principia Mathematica.*

GETTING THE ROYAL TREATMENT

Medical history is replete with horror stories of physicians administering bizarre and painful treatments to seriously ill patients in hope of curing ailments about which little was known. Consider what must be

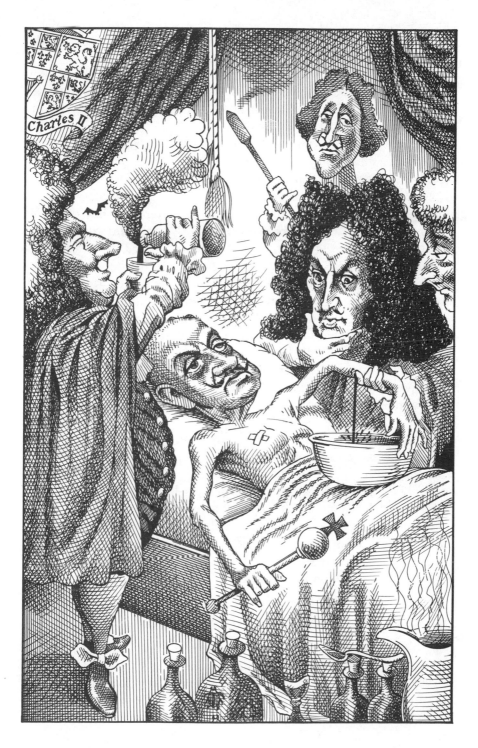

the classic medical horror story—the treatment accorded one of the kings of England during his dying days.

In 1685 Charles II fell ill of kidney disease and began to experience convulsions. For the next five days a parade of doctors—as many as twelve during the day and six at night—tried out their ideas on his progressively deteriorating body.

First they began by bleeding him, that favorite treatment of physicians of the time. They put cupping glasses on his shoulders, made incisions in his skin, and drew blood from his veins. Then they cut off his hair, put blistering agents on his scalp, and applied plasters of pitch and pigeon dung to the bottoms of his feet. After this they blew the offensive-smelling herb hellebore up his nostrils so that he would sneeze and thereby release the humors from his brain. After that they poured antimony and sulfate of zinc down his throat so that he would throw up and cleanse his insides. To clean out his bowels, they administered purgatives. And to stop the convulsions, they gave him spirit of human skull.

All the while this was going on, they periodically administered juleps for spasms, a gargle for sore throat, liquids to quench his thirst, tonics for his heart, and beer and broth for food. Sedatives, laxatives, herbs like cowslip and mint, even bezoar (a concoction found chiefly in the alimentary organs of cud-chewing animals and believed to be an antidote for poison)—all were given to the royal patient. As further precaution he was not allowed to sleep or talk. Not only physicians but priests, ministers, and servants entered the royal chamber to be of assistance. Finally, on the fifth day, the doctors drew twelve more ounces of the king's blood and gave him more heart tonics. This seemed to do the trick. By noon Charles II was dead.

In summing up the monarch's royal treatment at the hands of his physicians, Thomas Babington Macaulay, the nineteenth-century English historian and statesman, declared that the doctors had tortured the king like an Indian at the stake.

THE STOMACH BRUSH

The physicians of the Middle Ages knew little about anatomy. In fact, one of the physicians attending England's King Charles II, Gideon Harvey, wrote that "anatomy is no further necessary to a surgeon than the knowledge of the nature of wood to a carpenter or of stone to a stone cutter." Thus, some curious beliefs about the human insides emerged. One glaring misconception involved the stomach.

Physicians of the day viewed the stomach as little more than a basin with a waste pipe attached. It was therefore believed that a blockage could be unblocked upward by taking an emetic or downward by ingesting a purgative. In this regard, another of Charles II's physicians developed one of the most infernal of medical devices—the stomach brush.

The Astonishing History of the Medical Profession by E. S. Turner (Ballantine, 1961) gives a graphic picture of how the stomach brush worked:

> The idea was that the patient should drink a draught of warm water or spirits of wine in order to loosen the foulness from the stomach walls. The brush, moistened in some convenient liquor, was then to be introduced into the esophagus and slowly lowered into the stomach by twisting its wire handle. When it had reached its objective it was to be drawn up and down like a sucker in a syringe, the patient meanwhile drinking as much water as possible. A weekly use of the stomach brush was warranted to prolong life enormously.

One doctor advocated a substitute for the stomach brush. To get the same brushing action, he suggested drinking—twice daily—a glass of water containing fifty live millipedes.

THE FIGHT AGAINST SMALLPOX—AND AMONG THE FIGHTERS

One would think that Edward Jenner (1749–1823), an English physician who in 1796 performed the first successful vaccination against smallpox by using cowpox germs, would have gotten a warm welcome when he came forth—after eighteen years of study—with a way to protect people from one of the greatest scourges of human existence. But much of the medical fraternity of the day was less than enthusiastic. In fact, in 1798 when Jenner first publicly proposed vaccination with cowpox germs as a way to combat smallpox (inoculation with smallpox germs had been tried but was not safe), many doctors in England protested that the shots would give patients cowlike faces and would cause women to grow hairy and men to bellow like bulls.

Actually Jenner, who eventually gained wide renown for his discovery, was not the first to realize that cowpox could protect against smallpox. Two experimenters had preceded him, but unlike Jenner, both had failed to publish their results. Benjamin Jesty, a cattle breeder in the small English village of Yetminster, had been the first, as early as 1774, twenty years before Jenner. When he later realized his mistake of not publicizing his discovery and learned of the many awards being heaped on Jenner, he wrote letters to the Jennerian Society outlining his claim. He eventually was invited to London, but while Jenner got an honorary Doctor of Medicine degree from Oxford and 30,000 pounds from Parliament, Jesty wound up receiving a portrait of himself and a pair of gold-mounted needles.

THE ONCE MOST FIRMLY BELIEVED, BUT NOW WIDELY DISCREDITED, THEORY IN MEDICAL HISTORY

Masturbation is now considered by medical authorities a normal practice for sexual release, without leading to physical or emotional problems. But this thinking was not always the case. For centuries masturbation was viewed as the source of a whole series of ills.

One theory was that masturbation led to insanity. This concept began with the publication of *Onania, or the Heinous Sin of Self-Pollution,* published in 1716. The book was so popular it was translated into numerous languages and, in 1764, appeared in its eightieth edition.

Another widely disseminated belief was that masturbation caused homosexuality. By the end of the nineteenth century this was, in fact, standard psychiatric dogma.

But it was Dr. Benjamin Rush, writing in the first American textbook of psychiatry, published in 1812, who presented the true scope of thinking on the subject by most medical authorities of previous centuries. In a sentence that must be termed the summing-up on masturbation, Dr. Rush declared that the practice caused "seminal weakness, impotence, dysury [dysuria], tabes dorsalis, pulmonary consumption, dyspepsia, dimness of sight, vertigo, epilepsy, hypochondriasis, loss of memory, manalgia, fatuity and death."

THE BLEEDING OF LORD BYRON

Bleeding or bloodletting—the process of drawing blood from the body as a treatment for various diseases—was a popular medical practice for centuries. At one time it was seen as a way to remove bad humors and evil spirits from the body. Later it was viewed as a way to rid the body of bad blood. Today, however, it is rarely done. One of the few diseases in which bloodletting is still used is polycythemia; here the blood cells grow too quickly, and drawing blood helps remove excess amounts. But in earlier times doctors outdid themselves and freely prescribed the drawing of blood. They either placed three-inch-long bloodsucking leeches on the body or opened a vein. The procedure, however, usually only weakened the patient. Those doctors who were aggressive in the use of bloodletting often drained their patients of life.

A famous victim of bloodletting was the English romantic poet Lord Byron (1788–1824). In 1823 Byron, at the height of his considerable popularity, traveled to Greece to join the Greeks in their fight for independence from Turkey. He eventually contracted malaria and in April 1824 was attended by a group of physicians who agreed that as part of the needed treatment, Byron should be bled. At first Byron rejected the idea, but on April 16, 1824, the doctors pressed their case, saying that unless he agreed, they would not be responsible for "the consequences." According to Ethel Colburn Mayne's biography *Byron* (Charles Scribner's Sons, 1912), the still-reluctant poet eventually threw out his arms and angrily declared, "Come; you are, I see, a darned set of butchers. Take away as much blood as you will, but have done with it."

The first bloodletting did not help. The next day, April 17, Byron's condition worsened. This only spurred the doctors into renewed activity, and they bled him again—not once but twice more. Both times he fainted.

According to one account, the poet was eventually bled of "four pounds" of blood.

Byron finally died on April 19, 1824, three days after the doctors had begun bleeding him. He was thirty-six years old.

SURGEON LISTON WAS A REAL OPERATOR

The first operation performed in Europe with the aid of anesthesia occurred on December 21, 1846, at the University College Hospital of London. The surgeon was Robert Liston (1794–1847), and the surgery was the amputation of a leg.

The amputation, however, need not have occurred except for the mistake of the surgeon, who unwittingly had caused the patient's problem. The patient, a butler, had injured his tibia in a fall, and an infection had set in. But with pus beginning to drain from the wound, the body might very well have healed itself.

Surgeon Liston, however, examined the injury by making an incision, pushing his dirty finger into the opening, then feeling around the wound, thereby spreading the infection. Afterward he closed the incision, but then the signs of infection entering the bloodstream appeared: fever; sweating; headache; nausea; twitching; erratic pulse. The only action that could now be taken was to amputate the diseased leg; that is what the surgeon, who had caused the leg to become diseased, now did, thereby making medical history.

The surgeon who thus entered the history books had previously had other brushes with error in his career. Without the availability of anesthesia, surgeons had to operate quickly if the patient was to have any hopes of surviving the pain and trauma of an operation. Liston was widely known as being among the fastest with a scalpel. In fact, a legendary account of one of his especially fast amputations had him also accidentally cutting off two of his assistant's fingers and one of his patient's testicles.

THE DISEASE THAT DOCTORS CAUSED

Many diseases have been spread through the ignorance and mistakes of medical personnel. For instance, doctors became carriers of an ailment which killed thousands of women.

The disease was puerperal fever, a fatal infection that can suddenly attack a woman following childbirth. Its origin was unknown until the middle of the nineteenth century, when the causes were finally located. Until then a doctor could successfully deliver many babies, then watch helplessly as a string of new mothers he had attended died suddenly.

A doctor who realized that the problem rested with his colleagues was Ignaz Semmelweis (1818–1865). Hungarian-born Semmelweis received his medical degree in Vienna in 1844 and then became an assistant professor in the maternity department at the Vienna General Hospital, where 12 out of every 100 mothers were dying of puerperal fever—or childbed fever, as it was also called. After study and investigation he realized that the fever was contagious and soon declared that the doctors themselves were spreading the disease by not washing their hands thoroughly enough. He cited how medical students were bringing the disease from the dissecting rooms into the maternity wards and how a physician treating a woman with childbed fever would, with his unwashed or carelessly washed hands, pass the fever on to another patient. As further proof he pointed out that the incidence of the deadly fever was more common in the city's hospitals than in homes.

Many physicians strongly repudiated the idea that a doctor could be the culprit in the spread of the disease. Charles Meigs, professor of obstetrics at Jefferson Medical College, in Philadelphia, declared he could more easily attribute deaths from puerperal fever "to accident or Providence" than to "a contagion of which I cannot form any clear idea."

Semmelweis, however, was certain of his theory.* Starting in May, 1847, he began ordering doctors under him to wash their hands in chlorinated lime water. Childbed fever began dropping, from 12.24 percent to 3.04 percent at the end of 1847 and 1.27 percent by the end of 1848. While several university professors supported him, the head of the maternity department, Johann Klein, as well as other reactionary teachers, strongly opposed him. He was finally driven from his post in 1849, when a Hungarian revolt against Austria was overthrown and Austrian doctors could reject the Hungarian Semmelweis's authority—and his ideas.

Semmelweis went to Budapest where in 1850 he put his ideas into practice as an obstetric physician in the maternity department of a hospital there. Over the next six years, Semmelweis's antiseptic methods lowered the mortality rate in the maternity department to less than 1 percent. But back in Vienna, Austrian physicians went back to their old regimen, preferring to move from autopsy to patient and from one patient to another without worrying about the cleanliness of their hands. Childbed fever rose again in Vienna.

Semmelweis continued to push for the use of antiseptic methods in childbirth, even though he was often attacked and ridiculed by the medical fraternity. Ironically, he eventually became his own best evidence of the error of his colleagues' ways. In 1865, while treating

---

*Another who also noticed that medical personnel carried the contagion was Oliver Wendell Holmes, who wrote a treatise on the topic entitled *The Contagiousness of Puerperal Fever.* In it, Holmes cites the story of a physician who attended the postmortem on a woman who died of puerperal fever, then carried the diseased viscera in his pocket that day, and in the evening delivered a woman in labor without changing his clothes. She later died—as did the woman he attended the following day.

patients, he contracted childbed fever through a cut in his hand. By August, Semmelweis, then forty-seven, was himself dead of childbed fever.

At the time of his death the maligned pioneer of the use of antiseptic methods was in an insane asylum. There is some speculation that he, a bachelor, had contracted a venereal disease from a prostitute and the disease had caused his insanity. It is also conjectured that his two decades of waging the battle against fierce criticism had caused a mental breakdown. The year Semmelweis died was the year of British surgeon Joseph Lister's first antiseptic operation. Following that advance, the debate over Semmelweis's ideas ended. He had been proved right and his critics wrong.*

Medicine in the Modern Era

In modern times, even with the significant advances being made in the fight against disease and death, even with the improvements in diagnostic tools and the developments in medicines, the medical world must still combat its own infectious spread of error.

"TAKE TWO ASPIRINS, BUT DON'T CALL ME IN THE MORNING": A SURVEY OF MEDICAL MISDIAGNOSIS

How good are hospital staffs at diagnosing the ailments of patients? According to a recent study undertaken at a major Houston hospital, the record for diagnostic accuracy among interns and residents—the doctors who perform most of the patient care in a growing number of hospitals—is not very good at all.

Interns, who are medical school graduates in their first year of hospital training, were found to err by either missing a symptom or

---

*The mistakes about antiseptic conditions were also made on a wide scale, and also often with fatal results, when surgery became more prevalent following the introduction of anesthetics in 1826. While surgeons were becoming more skillful, infection was killing many of the patients now surviving the surgeries. At one point the mortality rate in surgeries reached 43 percent in an English hospital. In America it was 24 percent; in France, as high as 60 percent. One reason: During much of Queen Victoria's reign, the proper garb for a surgeon during an operation was a black frock coat. The coat, also donned during autopsies, was worn from one operation to another, with no thought given to cleaning the blood-caked garment. It is little wonder that the first medical man given a peerage (1897) was Lister, the surgeon who, like Semmelweis, advocated that cleanliness was next to successful medical care.

misinterpreting a symptom in 56 to 73 percent of their patients. Residents, who are in their second or third years or more of hospital training, in some cases had a worse result: They missed in 31 to 79 percent of their cases.

The study reported in the *Washington Post* was undertaken in 1982 at the 1,100-bed Houston Veterans Administration Hospital, which is affiliated with the Baylor University College of Medicine.

As part of the research 12 interns and residents examined 209 patients and made diagnoses. Eight senior staff doctors then repeated the exams and reviewed the initial diagnosis. The senior doctors found the interns and residents—referred to as house officers—had missed a total of 185 important symptoms and in seventy-one cases had made findings that had to be corrected.

Among the missed symptoms were such major items as "an enlarged liver or spleen or a cardiac murmur or a skin rash that could be a sign of cancer." Said Dr. Nelda Wray, one of those conducting the survey: "We're not talking about missing an ingrown toenail. . . . We're talking about findings that would significantly change and improve the treatment."

Dr. Wray told the *Washington Post* reporter that "this situation is common in American medical care." She pointed out that not only was this lapse inherent to those who make up hospital house staffs, but "Other studies have shown that older, more experienced doctors, even the senior staff at a Harvard teaching hospital, often do no better and sometimes do worse."

Dr. Wray placed part of the blame on the increasing reliance on machines to make the diagnosis and a shift away from the traditional physical exam. Her solution, spelled out in an article written with Dr. Joan Friedland in the *Journal of the American Medical Association* (February 25, 1983), calls for doctors to spend more time at bedsides and to pay more attention to the condition of the patient and his complaints. She also called for closer supervision of the doctor making the initial exam and diagnosis.

10 PERCENT DIE AFTER DIAGNOSTIC ERRORS

The conclusions of Dr. Wray and Dr. Friedland seem to be borne out by another study, this one a review of 100 autopsies conducted during the 1960, 1970, and 1980 academic years. In a report published in April

1983 in the *New England Journal of Medicine* by doctors at Brigham and Women's Hospital and Harvard Medical School, it was found that physicians missed the correct diagnosis in these deaths up to 24 percent of the time. In fact, one in ten of the patients studied might have been saved with the right diagnosis.

The findings and conclusions of this study paralleled those of Drs. Wray and Friedland. Overreliance on new technology and new procedures was leading to diagnostic mistakes. Singled out was the use of ultrasound, scanners, and computers which, the report said, "occasionally contributed directly to missed major diagnoses."

MEDICATION ERROR RATE: 12 PERCENT

Once a patient in a hospital gets a correct diagnosis, it seems he or she still has to worry that the correct medication will be given.

A study conducted in 1982 by the U.S. government's Health Care Financing Administration discovered that the medication error rate in surveyed hospitals averaged 12 percent. At one hospital, for instance, a doctor scribbled an order for Doriden, a sleeping compound. The nurse thought the word was Doxidan and gave that, instead of Doriden, to the patient. Doxidan is a laxative.

In another case a substitute nurse asked a patient, "Are you Mr. Thomas?"

The patient replied, "Wright."

The nurse, thinking she had found her man, gave Thomas's medication to Wright.

Mistakes in medication are widespread enough that Michael Cohen, director of pharmacy at Quakertown Community Hospital in Quakertown, Pennsylvania, and Neil Davis, a pharmacy professor at Temple University in Philadelphia, wrote a book on the subject—*Medication Errors: Causes and Prevention* (George F. Stickley Company, 1981). Their intention: to alert doctors, nurses, pharmacists, and other medical personnel to the problem.

While pointing out that "people are not dropping over every day from hospital errors," Cohen did acknowledge in an interview that "some result in serious complications, and in some cases deaths have occurred."

Cohen cited one such example: a child from Tennessee who died from too high a dose of a powerful anticancer drug. The wrong drug had been provided by the hospital's pharmacy.

AT LEAST THEY DON'T CHARGE FOR THEIR MISTAKES

One example of how mix-ups and mistakes occur in hospitals was reported by *Time* magazine in its March 31, 1980, issue. Entitled "Oops! A Horribly Regrettable Mix-up," the article told about two middle-aged women who underwent surgery the same morning at Philadelphia's Graduate Hospital. One needed an operation to remove the parathyroid glands in the front of her neck; the other needed cartilage removed from the vertebrae at the back of her neck.

Somehow, in a confusion the hospital was then still investigating, the surgeons mixed up the two women. The vertebrae operation was begun on the parathyroid patient; the parathyroid procedure, on the woman with the vertebrae problem.

Midway through the operations both doctors discovered their errors and stopped. Interestingly, while the operations were going on, each surgeon discovered something else wrong with his switched patient. The parathyroid of the vertebrae patient was found to have a benign nodule, which the doctor removed, and the parathyroid patient was cured of an unrelated pain in her leg when the doctor operating on her vertebrae somehow relieved pressure on her sciatic nerve.

Even with such beneficial, though totally unexpected, side benefits, the women were furious and considering lawsuits, and the hospital was calling the incident "horribly regrettable" and ordering an investigation. There was one further indication that the medical fraternity was taking this slipup very seriously. According to *Time,* as a gesture of goodwill, "neither the doctors nor the hospital will send either patient a bill for services rendered."

SURGEON, SPARE THY SCALPEL

On today's medical scene the public's greatest awe is probably reserved for the surgeon. The training, skill, and nerve necessary for a successful and productive career in surgery have placed the surgeon in a special pantheon in the medical profession.

This makes it all the more surprising to learn the high degree of operation-related errors. For instance, Dr. Robert G. Schneider, a Connecticut internist on the staff of Norwalk Hospital, a Yale-affiliated medical center, wrote a book about unnecessary, possibly harmful surgery entitled *When to Say "No" to Surgery* (Prentice Hall, 1982).

Because a friend of his had died while undergoing coronary artery bypass surgery, Dr. Schneider delved into medical literature, data from the National Institutes of Health (NIH) and the Center for Disease Control, and other scientific studies. He found some surprising figures:

—15 to 25 percent of the 20 to 25 million surgeries performed each year in the United States are unnecessary. And with some operations, the rate is even higher: for hysterectomies and tonsillectomies it is between 49 and 89 percent.

—40,000 to 80,000 deaths a year in America are caused by unnecessary surgical procedures.

—The fatality rate from general surgery is 1 out of every 78 patients.

—Fatalities from anesthesia alone occur in 1 out of every 1,500 operations.

Dr. Schneider cited another study that also indicates that many surgeries are unnecessary. A poll by *RN* magazine of 12,500 registered nurses working in hospitals found that half believed the number of unnecessary procedures stood at 30 percent.

The value of the increasingly popular coronary bypass surgery, according to Dr. Schneider, has never been proved in a statistically valid study. Yet the mortality rate, which averages 3 percent annually, has been as high as 7 percent in one year. The NIH, though, recently reported a no better return-to-work rate after bypass surgery than after more mundane medical treatment.

Some operations, although important for the relief of pain and for other reasons, show a higher mortality rate than the disease being treated surgically. A gallbladder operation, the fourth most often performed procedure in America, has a risk factor of 3 deaths in 1,000 operations—a greater risk to life than the disease itself poses.

Among the factors cited by Dr. Schneider as typical causes of death during surgery are cardiac arrest, infections, hemorrhage, pneumonia, and, of course, errors by surgeons and their staffs.

THE NOBEL PRIZE FOR AN OPERATION NOW SHUNNED

The prefrontal lobotomy is a surgical procedure developed in the 1940s in which parts of the front of the brain are removed to correct severe personality disorders. The problem with the operation is that patients undergo such dramatic personality changes that many become lethargic and infantile. For these reasons, this surgery is now widely discredited and rarely performed.

At one time, however, the lobotomy was so popular among surgeons that an estimated 50,000 people underwent the surgery in little more than a decade. In fact, in 1949, Antonio de Egas Moniz was awarded the Nobel Prize in medicine for originating the prefrontal lobotomy. Within ten years of the award, however, the operation was in disrepute. Still, thousands of people, many of them institutionalized, live today with lobotomies.

THE FAILURE RATE IS UP FOR A FAIL-SAFE OPERATION

One surgical procedure growing in use in recent years has been sterilization, primarily as a birth control measure. This procedure, too, has been plagued by an increasing number of errors and failures. In fact, lawsuits against physicians who perform unsuccessful sterilizations have also been increasing.

Dr. Philip Darney, a doctor on the staff of the obstetrics and gynecology department at San Francisco General Hospital, told a Planned Parenthood convention about an "outbreak" of sterilization failures at his hospital between January 1979 and June 1981.

During that period 13 of the 817 women who had had sterilization procedures later became pregnant—a failure rate of 1.6 percent. Dr. Darney noted that the normal failure rate is approximately 5 out of every 1,000 operations—or one-half of 1 percent. (Although this may be a small percentage, it is still an alarming figure for a procedure popularly perceived as offering foolproof contraception.)

The failure rate went up if cautery, one of several methods of tubal sterilization, was used. At Dr. Darney's hospital the failure rate reached 2.6 percent. And when relatively inexperienced surgical residents did the operation, the failures zoomed to 10 percent or higher.

Interestingly, at the same convention another doctor, Richard M. Soderstrom of Seattle, told the group of family planning professionals that women who become pregnant after undergoing a sterilization procedure are—as he put it—"particularly vulnerable to suggestions of malpractice litigation."

Among his suggestions for heading off litigation, Dr. Soderstrom said that the doctor who performed the unsuccessful sterilization should perform the operation again for free.

CAN WE HAVE THE NAME OF THAT DOCTOR, PLEASE?

In Southampton, England, in 1967 a woman gave birth in the same hospital in which she had been sterilized sixteen months before.

HEARTTHROB: THE INCORRECT USE OF PACEMAKERS

Another procedure being found to have incorrect applications is the implantation of pacemakers to control the heartbeat in patients with cardiac problems. A four-year study conducted in a New York City hospital and published in the *Journal of the American Medical Association* on August 14, 1981, revealed that many pacemakers are being implanted in people who don't really need them.

Dr. Howard S. Friedman, a researcher at Brooklyn Hospital and author of the article, stated, "A lot of pacemakers are being implanted in patients who shouldn't have them. Physicians need to be more critical about making a diagnosis that indicates implanting a pacemaker. If they did this, they would implant fewer of them."

Pacemakers are small electronic devices placed surgically in the body and connected to the heart. They work by providing mild electric shocks to stimulate heart muscles and thereby make the heart beat regularly. From 1973 through 1978 the number of pacemakers in use in the United States increased by about 9,000 a year, to an estimated 85,000 in 1980, the year before Dr. Friedman's study was published.

When a physicians' committee was set up in Brooklyn Hospital in 1977 to review the use of pacemakers, they found the number of such procedures dropped from forty-eight in 1976 to twenty-two in 1977. Also, the committee found that before the peer review 57 percent of the patients receiving pacemakers were alive three years after the operation, but after the peer review the figure jumped to 87 percent. Dr. Friedman's conclusion: The 30 percent jump in the pacemaker effectiveness showed that previously pacemakers had been given incorrectly to patients for whom the device had no value.

Some overuse of pacemakers resulted from lapses in medical knowledge at the time. In eight cases, for example, it was found that patients taking a drug to lower blood pressure were provided with pacemakers because their heart rhythm had slowed; only afterward did doctors learn that the drug these patients had been taking to lower blood pressure also produced the slow heart rhythm.

The Brooklyn Hospital study was not the first to find a misuse of pacemakers. Several years before, research carried out at the Harvard Medical School discovered that ten of the thirty-two patients studied at the time did not need pacemakers either.

EVEN A FIRST LADY NEEDS A SECOND OPINION

Even the wealthy and influential are not immune from experiencing the results of medical mistake.

Eleanor Roosevelt, wife of President Franklin Delano Roosevelt, lived to the age of seventy-eight, but she might have lived longer if not for her doctors' errors in diagnosing her ailment. In fact, her situation at death was for a long time hushed up and covered over; writing about it in the Baltimore *Sun* of June 28, 1981, two doctors referred to it as a "case that has been whispered about within medical circles since her death some 20 years ago."

Eleanor Roosevelt's problems began when she became ill and started bleeding and bruising at the slightest touch. Her doctors agreed on a somber diagnosis: The widow of a U.S. president was suffering from a fatal blood disease. They decided to forgo one crucial but painful test—an open bone marrow biopsy—to confirm this conclusion and instead began treating her with prednisone. This drug, usually helpful in fighting blood diseases, has one serious side effect: It lowers the body's resistance to infection.

After Eleanor Roosevelt took the drug, her condition rapidly worsened, and she soon died. Only later did her physicians learn that their diagnosis had been wrong. The former first lady was suffering not from a blood disease but from tuberculosis. The biopsy would have told her doctors they had erred. But their treatment was an even worse mistake. The prednisone, because it lowered her resistance to infection, actually enabled the tuberculosis to spread even faster. Not only had the medical evaluation been wrong, but the prescribed medication had been the worst possible for Eleanor Roosevelt's real malady.

79

ELEANOR ROOSEVELT AND HER DOCTORS
(continued)

Eleanor Roosevelt seems to have had her problems with doctors. One doctor suggested that to improve her memory, she should eat garlic. Ever willing to follow doctors' orders, Mrs. Roosevelt thereafter ate three chocolate-covered garlic balls every morning for the rest of her life.

GREAT MOMENTS IN MEDICINE

Dr. Harvey Cushing was a renowned surgeon at the Johns Hopkins Hospital in the early 1900s. Among his achievements he perfected the techniques of neurosurgery (in his career he removed more than 2,000 brain tumors). For twenty years he also did work on the pituitary gland, which led him to describe a new disorder, termed Cushing's syndrome. On February 29, 1932, he delivered to the Johns Hopkins Medical Society his first official presentation on his findings about how the pituitary worked as a master gland influencing the other endocrine glands: the thyroid, the sex glands, and the adrenals. It was here that Cushing labeled the signs and symptoms characterizing a particular disease entity as syndromes, at the time an unfamiliar term. Throughout his speech Cushing mispronounced *syndrome,* giving it three syllables and accenting the third. One of those in the prestigious audience that day was Dr. William Welch, a fellow member of the Hopkins staff and a noted pathologist (he taught the first pathology course in America). After the presentation, with everyone eagerly awaiting his reaction, Welch came to the podium and talked not about Dr. Cushing's findings but about the derivation of the word *syndrome.* For forty minutes Welch went on about the origin and history of the word, concluding by telling Cushing that he had mispronounced the word and that the accent was on the first syllable.

* * *

Dr. William Halsted, one of the first surgeons to understand the need for antiseptic conditions during surgery to prevent infection, added his own contribution to cleanliness in the operating room—rubber gloves. One of the first surgeries in which sterilized rubber gloves were used by Dr. Halsted and his entire operating team occurred in 1895 and involved the operation on the compound fracture of a kneecap. In the past such a surgery usually proved unsuccessful, leading to amputation of the leg, because infection set in quickly after the knee had been opened. To guard against this possibility, Halsted had the area on the patient meticulously sterilized over three days. As the operation was about to begin, Halsted turned to the physician who had brought the patient in and asked him how far apart had been the two fragments of the fractured kneecap. The old, bearded doctor responded, "The upper fragment was right here," and put his dirty hand on the kneecap that Halsted's surgical team had spent three days cleaning. Halsted cried out at the sight of the grimy hand touching the patient, and the entire cleaning-up procedure had to be repeated.

* * *

Johann von Mikulicz-Radecki, of Poland, became one of the most famous surgeons of Europe in the late 1890s. Doctors came from far and wide to watch him operate. Mikulicz, however, had one problem. He was not very interested in clothes or conscious of how he dressed. During one operation, with a number of fellow doctors looking on, the renowned surgeon's pants fell down.

NOW HEAR THIS!

In the 1950s Harold Senby of Leeds, England, began experiencing difficulty hearing and was fitted for a hearing aid. He noticed afterward that his hearing did not improve. In fact, he seemed to be hearing less than before. During the next twenty years he was refitted several times, but each time his hearing stayed the same.

Finally, during a visit to his doctor in March 1978, the hearing aid was removed entirely, and Senby began to experience improved hearing. After a closer medical examination it was then discovered that in the 1950s the hearing aid mold had been made for his left ear, not his right ear, which was the one giving him trouble. Now, with the hearing aid removed, he was finding he had "almost perfect hearing."

Senby, in later commenting on how his doctors could have mishandled his hearing problem for twenty years, said, "Over the years I have been fitted with several new aids, but no one noticed that I had been wearing them in the wrong hole."

THE POSTER CHILD WHO NEVER HAD THE DISEASE

Cystic fibrosis is an incurable, often fatal respiratory illness that afflicts children. Most who contract it never survive their mid-teens.

Rodney Brown of Indiana was fifteen months old in 1970 when doctors first diagnosed his pneumatic condition as stemming from cystic fibrosis and told his parents that their young son had the dreaded disease. On the basis of that diagnosis, every day the child was fed numerous antibiotics and taken for physical and inhalation therapy to help him with his breathing. Every six months he was also subjected to X rays of his lungs.

This went on for ten years. During that time, Rodney was chosen, in 1975, as Indiana's cystic fibrosis poster child and, in 1976, as the national poster child. In the latter year, as the Cystic Fibrosis Foundation poster child, he met and posed with President Gerald Ford.

Then, in 1980 the family moved to Maryland, and soon a new specialist treating Rodney conducted some further tests. The first test, conducted at the Johns Hopkins Hospital in Baltimore, showed a surprising result: Rodney did not have cystic fibrosis. He was then retested at Children's Hospital in Washington, D.C., and tested a third time at the National Institutes of Health in Bethesda, Maryland, outside the nation's capital. All three tests confirmed the same finding: The original diagnosis had been a mistake.

The misdiagnosis came to public light only in 1982, when a reporter from Indiana who had written about Rodney asked the local Cystic Fibrosis Foundation chapter how he was faring since Rodney was now entering his teens. It was then that foundation officials related the story of the faulty diagnosis.

According to his parents, when Rodney had finally been told that he did not have the fatal disease but had only an asthmalike condition, the youngster's first reaction was "I got to see the president for free!"

THE FORTY-EIGHT-YEAR MISTAKE

Catherine Yasinchuk, who died in February 1983 at the age of eighty-six, spent forty-eight years of her life in a Pennsylvania mental institution because officials did not realize the only language she could speak was Ukrainian.

Catherine first came to the United States several years before World War I at the age of fifteen. She was alone, had no family or friends, and did not know English. Eventually she married and had a child. When first her child and then her husband died, she became hysterical. Police found her wandering the streets, and when they tried to question her, she did not seem to understand. Police said she only babbled.

Unable to communicate with her, authorities institutionalized her in a mental health hospital. She was twenty-three. For six years she continued to babble and did not talk to doctors and staff. Then, in 1927, she stopped talking altogether. She just sat, walked around, and stared at the walls.

This was her situation for the next forty-one years, until 1968, when a new director of the Philadelphia State Hospital at Byberry, Pennsylvania, declared that the institution would not be a "dumping ground" for unwanted individuals. Miss Yasinchuk's case then came up for review, and the staff tried once again to communicate with her. Russian, German, Austrian dialects were tried. Also Polish and Lithuanian.

Then an employee in the hospital's personnel department tried Ukrainian. Miss Yasinchuk began to respond and slowly started to talk.

A year later, in December 1969, at the age of seventy-two, Miss Yasinchuk was released from the mental institution she had entered forty-eight years before. Her guardian was the daughter of the woman in the personnel department who had first communicated with her in Ukrainian.

Catherine Yasinchuk moved to a home for the elderly, adjusted slowly to the world outside the mental hospital, and lived in the home until 1980, when she broke her hip. She was then moved to a nursing home in the suburbs of Philadelphia, where she passed the last three years of her life. But because of the failure of health officials to communicate with her, more than half her long life had been spent unnecessarily in a mental institution.

WHEN THE PATIENT ERRED

Sometimes the patient flouts the doctor's advice and winds up making a fatal error. Consider the case of fifty-four-year-old Chao Boonchu, of Thailand.

Boonchu proclaimed himself his country's champion chain smoker. In 1981 the Thai press featured stories about him and his boast that he

THE BABYLONIANS HAD AN IDEA

In the modern world, the only punishment for the doctor who messes up is a malpractice suit and possibly being stripped of his license to practice. The Babylonian Code of Hammurabi, written about 5000 years ago in the cradle of civilization, has the first known penalty for errant doctors. If a patient died while an abscess was being opened and drained, the physician was to be disciplined with what we would consider a slightly harsh punishment. The doctor was to have his hands cut off.

held an unbreakable record: He proudly claimed to have smoked 120 cigarettes a day for thirty years with no ill effects.

Within a year, however, it seems that the error of Boonchu's ways became apparent. In August 1982 the middle-aged man collapsed with severe breathing difficulties and was rushed to a Bangkok hospital, suffering from severe heart and lung complaints. He was last reported lying in a coma in critical condition.

OUR FAVORITE MISDIAGNOSIS

Ted Crail, an author of books on animals and wildlife, became interested in his life's work when as a six-year-old he was told he was dying of Bright's disease and that there was "no chance" that he could live until Christmas. As a result, his family celebrated Christmas in September for him and, as a present, took him on a tour of the wilderness and wildlife in Montana.

But Crail did not die. Decades later he was still living—and writing in one of his books, *Apetalk & Whalespeak* (J.P. Tarcher, 1981), about how, because of the misdiagnosis, "I was left with a distrust of those who insist they know 'the scientific facts' and are very peremptory about it. . . ."

Although Crail did not pass away when he was six years old, someone else soon did. Who? The doctor who had told him he had less than a year to live.

Crail writes: "I have been declared dying a time or two since and it always seems to be the doctor who dies."

THE "WATCH OUT FOR EXPERTS" DEPARTMENT:

SOME OFFSPRING OF
THE FATHER OF MEDICINE

The Greek physician Hippocrates (460?–377? B.C.) is often called the father of medicine. His concepts of healing were far advanced for his time, and he helped elevate medical care out of superstition and into the realm of rationalism. Even today doctors take the Hippocratic Oath, swearing allegiance to a code of medical ethics first developed by Hippocrates.

But Hippocrates was not above making glaring mistakes in his theories about health and the practice of medicine. Here, gleaned from his book *The Aphorisms* and other writings, are some Hippocratic errors:

—He said that one could tell the sex of an unborn child by which of the mother's breasts became larger and on which side of the womb the fetus developed.

—If a woman did not conceive but wanted to know if she was capable of becoming pregnant, she should be covered with wraps and perfumes burned underneath her. Hippocrates wrote: "If the smell seems to pass through the body to the mouth and nostrils, be assured that the woman is not barren through her own physical fault."

—He had good news for people subject to jaundice: They are not very susceptible to flatulence.

—He had bad news for people with speech impediments: They are very likely to get protracted diarrhea.

—He had good and bad news for the young: They can get gout only after sexual intercourse.

—South winds, according to Hippocrates, cause deafness, dimness of vision, heaviness of the head, and torpor. A north wind causes coughs, sore throats, constipation, shivering, and pains in the sides and chest.

—And the father of medicine even had an erroneous message of hope for those without hair: Bald people who get varicose veins, he wrote, grow hair again.

CHAPTER III

Flub:
Error in Science

*For every human problem, there is a solution
that is simple, neat . . . and wrong.*

—H. L. Mencken

The human race has tried to gain control of its environment through science. By developing, testing, and implementing theories for why solids, liquids, and gases behave the way they do—why the physical and chemical properties of our universe are the way they are—people have hoped to predict and thereby have power over their world.

But the advances in science have not been smooth or sure. About 600 B.C. the Greeks began the first systematic look at the world to discover why it operated in the way they saw it operated. So influential were their ideas and pronouncements that, as one history of science has noted, "those theories took the whole world captive, and for twenty-two centuries, Western science was Greek science; for the Romans, the

Arabs and the men of medieval Europe did no more than enlarge on Greek ideas."

The only problem was that while science was virtually all Greek to the world, Greek science itself was more often than not in error. The passage of time showed only too well that "the Greeks reasoned better than they observed, so that, too often, they based beautiful theories on unsound data."

By the end of the Middle Ages people had begun developing the methods and tools to make the observations and to conduct the experiments that would lead to valid scientific conclusions. The last 300 years have seen the most rapid and wide-sweeping advances in science, a time that has paralleled medicine's advances as well.

But the human element has never been removed from the scientific element, and from ancient to modern times human error has often been present to cloud progress and delay advances—although one can't help feeling that mistakes have been a necessary part of the march of science, too.

Consider, for instance, the case of the English statesman, philosopher, and scientist, Francis Bacon (1561–1626). He was one of the first proponents of empiricism—of solving scientific problems through experiment. He thought all previous claims to knowledge by medieval science to be wrong, and he campaigned vigorously for the investigation of nature by collecting as many facts as possible.

Except that Bacon possibly went too far for his own good.

To determine if cold could slow down decay, he stuffed a dead chicken with snow, but contracted a chill while performing the experiment. After several days, the father of scientific empiricism died of bronchitis.

As far as is known, this was the only experiment Francis Bacon ever attempted.

The Great March of Science

THE 1,500-YEAR ERROR: PTOLEMY'S THEORY OF AN EARTH-CENTERED UNIVERSE

About A.D. 150 an astronomer working in Alexandria, Egypt, recorded his observations and theories about how the earth, sun, and planets work in relationship to one another. His efforts filled thirteen volumes and made such an impact on his generation and those that followed that the work—entitled *Mathematike Syntaxis* or *Mathematical Composition*—came to be known as the *Almagest,* a Greek-Arabic term meaning "the greatest."

The author was Claudius Ptolemy, who was born in Greece but about whom little is known except for his theories. And those theories held such sway that they were believed and followed for 1,500 years—even though many of them were wrong. (Indeed, Ptolemy's erroneous theories were based on very much the same observations made by scientists centuries later—yet they came to more accurate conclusions.)

For example:

- He rejected the concept that the earth moves (he pointed out that if it did move, animals, people and things would be thrown into the air).
- He said that the fixed, motionless earth was located at the center of the universe.
- He believed that the moon, sun, and planets moved around the earth at different speeds.
- He stated that the stars, very bright spots of light, sat in a concave dome that arched over the universe.

Ptolemy was also celebrated as a geographer, but here too his work is, according to the *Encyclopaedia Britannica,* "disfigured by some astounding errors." He accepted an estimate of the size of the earth that was incorrect, so his map of the planet was greatly skewed—"an error vitiating all his conclusions." He "inexplicably" followed the Greek philosopher Posidonius's estimate of the circumference of the earth as 18,000 miles, rather than the correct figure of about 25,000 miles—which Eratosthenes had estimated a hundred years before in 240 B.C. (One possible reason why Ptolemy chose the smaller figure was that the larger figure would have meant that the known world only occupied one-fourth of the earth's surface, a truth Ptolemy found difficult to accept.)

Thus, Ptolemaic maps exaggerated the landmass between Spain and China and underestimated the width of the ocean dividing them, showing Asia and Europe to be only 4,000 miles apart, rather than the correct 12,000 miles. One inadvertent benefit of Ptolemy's error: These incorrect maps persuaded Columbus of the feasibility of sailing from Europe to Asia.

Ptolemy also dealt seriously with astrology, thereby helping spread what was a superstition and giving it credence in many minds as a science.

These, then, are some of the ideas of the man whose system of astronomy was accepted as authoritative until 1543, when the Polish astronomer Copernicus finally proved that Ptolemy's major concepts had been wrong all along.*

But in the interval, for those 1,500 years, Ptolemy's errors delayed or distorted other scientific inquiries.

ARISTOTLE'S MISTAKEN LAW: THE BIGGER THEY ARE, THE HARDER THEY DON'T NECESSARILY FALL

Aristotle is one of those towering figures of history, a man revered as philosopher and scientist. But he fouled up when it came to determining one of the basic scientific principles: the property of falling bodies.

Do or don't things fall at the same rate, no matter what their weight?

The problem seems to offer a very simple way to find out. Just drop balls or rocks of different weights from atop a building and see if they hit the ground at the same time. Simple, right?

Well, for some reason this procedure eluded Aristotle. And it took people about eighteen centuries to sort out the truth.

Aristotle came to the conclusion that the heavier a body was, the faster it fell. He also added that the density of what it fell through determined the rate at which it fell.

To the naked eye—and the unclothed mind—this would seem to be a true and accurate law of science. Except for two things: Number one, there is no record that Aristotle made any measurements to corroborate this. Number two, he also did not take any account of or understand the concept of the acceleration of falling bodies. He therefore came up with two delightful but erroneous laws:

1. A body falls at a speed in proportion to its weight (in other words, according to Aristotle, a ten-pound weight falls ten times faster than a one-pound weight).
2. The speed of a falling body is inversely proportional to the resistance of what it falls through.

*Not all of Copernicus's ideas were correct either. For instance, he stated that the orbits of the planets around the sun are perfect circles. However, the orbits are elliptical.

The truth is that a ten-pound rock and a one-pound rock fall at the same rate and, dropped from equal heights, hit the ground at virtually the same time. Also, a stone falling through a vacuum would encounter no resistance and therefore, according to Aristotle's thinking, would fall at an infinite velocity, but Aristotle, saying such a possibility was absurd, therefore concluded that a vacuum could not exist.

Even though Aristotle's two laws were wrong and led to a tangled web of thinking to make them seem to work, no one appears to have contested these concepts or to have tested the Greek philosopher's conclusions about falling bodies.

In fact, no record exists of any such tests until the sixteenth century, when Simon Stevin, a Dutch mathematician, conducted such an experiment and wrote about it in a book. He disproved Aristotle's theory by simply dropping two leaden balls—one ten times heavier than the other —from a height of thirty feet onto a board on the ground and observing that "the lightest does not take ten times longer to fall than the heaviest, but that they fall so equally upon the board that both noises appear as a single sensation of sound."

Stevin wrote his book in Dutch, which was a language not read by many people in Europe, so his findings went neglected for fifty years, until Galileo, who does not appear to have known of Stevin's findings, showed with his experiment that Aristotle had erred.

Galileo (1564–1642) tested Aristotle's laws by running balls down an inclined beam since rolling became identical with falling as the beam was raised to the vertical. Ironically, while Galileo set out to disprove an error in Aristotle's thinking, his attempt gave rise to an error in popular history—namely, that he conducted these experiments by dropping weights from the top of the Leaning Tower of Pisa (there is no such record in his writings or in anyone else's of the time).

Galileo's demonstration of Aristotle's 2,000-year-old error about the speed of falling bodies had one other important, albeit negative, result for the history of science. He antagonized other scientists and made himself a controversial figure by challenging and disproving Aristotle. Later, when Galileo tried to show that another Greek scientific thinker, Ptolemy, had been wrong about an earth-centered universe and that the sixteenth-century scientist Copernicus was right to oppose this concept, church authorities had the support of scientists of the day in stilling Galileo and getting him to recant his challenge of Ptolemy.

THE QUESTION IS: DID THE APPLE HIT NEWTON ON THE HEAD?

Isaac Newton, the scientist who made giant discoveries because, as he said, he stood on the shoulders of giants, almost made some giant mistakes. At the very least, he wasted a lot of time on nonscientific matters he thought were scientific.

Newton (1642–1727) was the most celebrated scientist of his day and one of the most original scientific thinkers in history. He developed laws of motion and optics, described gravity, helped invent calculus, and set the stage for modern physics.

This same scientist, however, also spent twenty-five years after his monumental early discoveries delving into alchemy in a quest to find magical potions and to gain occult powers. Performing most of his work on alchemy in secret, he amassed 100 volumes and eventually wrote 650,000 words on the subject.

Think what those years of work and writing could have revealed about the physical world if Newton had not spent so much effort on the dead end of alchemy but instead had devoted himself to scientific matters. After all, not one significant finding emerged from all his mixing of metals and heating of elements. Interestingly, while he worked in secret on his alchemy, Newton was also serving as president of the world's most prestigious scientific body, the Royal Society.*

WHY THE STORY OF FRANKLIN'S LIGHTNING ROD HAS THE WRONG ENDING

Benjamin Franklin is famous for many scientific experiments and discoveries, especially his work with electricity and lightning. The man who flew a kite and found that lightning is electricity went on to invent

---

*Newton was not the only great scientist to try to employ a less than scientific approach to solving the mysteries of the physical world. Alfred Russel Wallace, who also developed the theory of evolution at the same time as Charles Darwin, had made a name for himself as a naturalist, with many books and papers to his credit. But Wallace also believed that spirits were active in the natural world. He participated in séances and tried to communicate with the dead. He also wrote a number of papers on the subject as a way to explain one of the imponderables of evolution: how human beings had descended from the apes. His answer: The spirit world had intervened and made it possible.

the lightning rod as a way to protect buildings from being struck. He was, however, wrong about one important aspect of his lightning rod.

Franklin advocated a sharp-pointed rod as the best protection against the electrical discharge of a lightning bolt. Interestingly, King George III, a science buff, disagreed. He said that a blunt-edged rod would work better. While a war would later be fought between the Americans and the British over more monumental matters, the people of the two countries went their separate ways over this issue, too.

Two hundred years later scientists discovered what makes the best lightning rod. Their conclusion: Franklin was wrong; King George III, right.

In an article in the science section of the *New York Times* of June 14, 1983 ("Lightning Rods: Franklin Had It Wrong"), science writer John Noble Wilford reported on the findings of Dr. Charles B. Moore, a New Mexico physicist, who experimented and found that the electric fields above blunt-edged rods were as much as two times stronger over greater distances than those above sharper-edged rods, a significant factor in protecting against the destruction of lightning.

In fact, Franklin's sharp-pointed rod was found to create around its tip "a dense sheath of electrified, or ionized, particles which reduce the probability of lightning's striking the rod." In other words, because Franklin's lightning rod does not readily draw lightning to it, it is actually less likely than King George's to divert lightning from striking nearby structures.

Franklin had originally erred in another way about lightning rods. He first saw them as a way to prevent lightning from even forming. He thought that the electric charge in thunderstorms could be conducted away from them by a pointed iron rod, raised above the ground but connected to it by a wire. Only later did he realize the true way in which a lightning rod could function—by carrying the lightning discharge harmlessly into the ground.*

---

*Actually it is a mistake, based on an illusion, to think that lightning strikes downward. It really strikes upward from the ground. The flash of crackling light comes from upward surges of electricity being discharged by the electric fields of objects on earth. These discharges are the completions of the circuit produced by the downward flow of electric charges coming from rising warm, moist air mixing with colder air in thunderclouds. Without a lightning rod and its upward currents to hook up with this electricity and ground it, these downward currents might have hooked up with the discharges from the electric fields given off by other objects on the ground. Since these are not grounded, the destruction of property or the death of people can occur.

WHERE DOES LIGHTNING STRIKE TWENTY-THREE TIMES IN ONE PLACE?

One of the mistaken notions about our physical world is that lightning does not strike twice in one place. In any one day 2,000 thunderstorms occur around the world, and many of them give off lightning, some undoubtedly hitting places or objects more than once. In New York City alone one spot is struck by lightning nearly two dozen times a year. It's the Empire State Building. Because the radio-TV antenna atop that structure acts as a lightning rod, the skyscraper is hit by lightning an average of twenty-three times annually.

HOW POLITICS KILLED A GREAT SCIENTIFIC CAREER

Antoine Lavoisier (1743–1794), a brilliant French scientist, was the first to describe the process of chemical combination. However, his great contributions to science were cut short when, at the height of his career, he was executed because of his political sympathies. During the French Revolution a judge ordered him to the guillotine, saying flippantly the Republic had no need for scientists. Lavoisier is said to have responded, "This probably saves me from the inconvenience of old age."

THINGS THAT GO BUMP ON THE HEAD: THE THEORY OF PHRENOLOGY

Franz J. Gall (1758–1828), a scientist and physician who made a special study of the brain, claimed that characteristics of the mind were related to the physical characteristics of the skull. He called this approach phrenology and devised an elaborate chart showing the relationship of emotion, talent, and ability to various parts of the head. The idea was that a prominence or bump in one part of the skull indicated a definite characteristic of the person.

In what must have been a picturesque procession, Gall went around feeling the heads of students, men, women, children, even convicts. He claimed that he could determine aspects about a person by palpating his head. For instance, he said he could feel the "organ" of number in mathematicians, tune in musicians, language in poets. He even claimed

93

that certain bumps made people criminals and that the insane had heads of a certain shape.

Phrenology became popular at the turn of the nineteenth century, but the scientific basis for the theory has never been shown. The shape of the brain cannot be determined by the shape of the skull. Although different parts of the brain perform different functions, dividing the brain into the different characteristics of people is unscientific. Despite the popular interest in it, phrenology has been labeled at best a pseudoscience. The term now usually applied to phrenology is *quackery.*

One real victim of phrenology may have been Gall himself. He had previously made important discoveries about the brain and could have been celebrated as a pioneer in brain physiology. Instead, he is now considered just a strange footnote in the history of science.

THE ASTRONOMER WHO DISCOVERED A PLANET THAT DOES NOT EXIST

Urbain Jean Joseph Leverrier was an astronomer who discovered two planets, one of which never existed.

In 1846, after studying irregularities in the orbit of the planet Uranus, Leverrier calculated that a hitherto unknown planet was exerting its gravity on Uranus. After he had predicted the position at which this planet would be found, another astronomer, following the calculations, confirmed that Leverrier was indeed correct. Thus was the planet Neptune discovered.

The news astonished the scientific world. Leverrier found himself celebrated. He was named director of the Paris Observatory, and a chair of celestial mechanics was created for him at the Sorbonne.*

Heady with his newfound fame and in awe of his own technique for finding planets, Leverrier next studied the orbit of Mercury and soon announced that it, too, moved as if an unknown planet were affecting its orbit. An amateur astronomer then announced some sightings of a possible planet. This was enough for Leverrier. In 1859 he announced his discovery of that very planet. He called it Vulcan.

Leverrier's announcement created another international stir. "All astronomers of all countries will unite in applauding this second triumphant conclusion to the theoretical inquiries of M. Leverrier," the

---

*When William Herschel discovered Uranus in 1781, he, too, became famous almost overnight. So unknown had he been that various journals misspelled his name as "Mersthel," "Herthel," and "Hermstel."

Royal Astronomical Society editorialized in its bulletin about the astronomer who had now discovered two planets.

This time, though, the acclaim that began mounting for Leverrier soon slowed and then stopped because no one could confirm his findings. No professional astronomer could now locate the planet Vulcan. Soon it became learned that Leverrier had not had the Paris Observatory check out his calculations or the sightings of the amateur astronomer. When sighting of the planet or corroboration of its existence continued to elude scientists, people became disenchanted with Leverrier. However, he died in 1877 still insisting on the existence of Vulcan.

The question about the glitch in Mercury's orbit was not resolved until the next century. We now know that Mercury's orbit is not affected by another planet but can be explained by Einstein's theory of relativity, which indicates that the existence of the sun's massive gravity affects the way in which nearby Mercury travels through space.

THOSE MARTIAN CANALS

In 1877 a respected astronomer, the Italian G. V. Schiaparelli, looking through his telescope under unusually favorable conditions for viewing Mars, saw streaks crisscrossing the planet. His announcement of his finding these hitherto unseen lines across the face of Mars was taken seriously by other astronomers, some of whom, upon looking through their telescopes, now said they could see the streaks, too.

The problem was that his announcement was also taken too literally, for Schiaparelli called the streaks *canali,* an Italian word which means "channels" or "grooves" and which connotes features made by nature. Schiaparelli never said that the *canali* were constructed by intelligent beings. But the word was mistranslated into English, emerging as *canals,* which many people interpreted as meaning an artificially built structure.

The English word gained wide currency, and soon speculation spread among the public that intelligent beings had constructed the canals.

Such an idea received a boost from the American astronomer Percival Lowell. As developer of the Lowell Observatory in Flagstaff, Arizona, considered the best equipped in the world, Lowell was able to view Mars more clearly than anyone else. Peering through his powerful telescope, he saw not only the many straight-lined canals but dark spots where they crossed each other. He termed such spots oases and announced his theory: The canals had been built by intelligent beings to transport water from the melting polar icecaps of Mars to irrigate the planet's arid parts.

Not all astronomers supported Lowell—or even agreed that canals existed on Mars. Differences in telescopes and in atmospheric conditions could make viewing Mars difficult, and to many people the canals either looked too faint for study or could not be seen. Even photographs of the planet did not show features well.

When Lowell died in 1916, still a strong supporter of the canal theory, the controversy was left in limbo. There was no way to prove or disprove the idea until the 1970s, when photographs taken first by space probes and then by a TV camera placed on the surface of Mars failed to show any canals. And even though cracks were discovered in the Martian surface, implying the existence at one time of rivers, these cracks proved too faint to have shown up in the telescopes of Schiaparelli, Lowell, and others.

The conclusion now is that the canals, which had been the basis for so much erroneous theorizing about life on Mars, were an optical illusion.

SO WHAT'S A FEW PERCENTAGE POINTS?

For many years it was believed that Pluto's mass was more than 90 percent of the Earth's. Recent calculations . . . however suggest that Pluto's mass is less than 10 percent of the Earth's.

—*CBS News Almanac*

FREUD'S SLIP

Sigmund Freud, searching for a substance that would do wonders for the brain, came across cocaine, ingested it, and found it to be a "magical drug."

Without fully understanding its properties or running controlled tests, an enthusiastic Freud began advocating its use by others—patients, colleagues, family, and friends.

He became such a vigorous proselytizer for cocaine that he wrote it could treat indigestion, vomiting, depression, anemia, asthma, and the withdrawal symptoms of morphine and alcohol addiction. He saw cocaine as a stimulant, an anesthetic, an aphrodisiac, and more.

Eventually Freud's enthusiasm for the drug led to its widespread adoption before its bad side effects became apparent. In the meantime, a close friend of his died from the drug.

Freud finally stopped using cocaine, but he gained for himself the reputation as the man who had introduced to Europe "the third scourge of humanity." Only alcohol and morphine were seen as greater threats.

A TREE GROWS IN MOSCOW—BARELY: LYSENKO'S THEORIES OF GENETICS

Trofim Denisovich Lysenko, the son of Russian farmers, was the most important biologist in Russia from the 1940s until the 1960s. Both Joseph Stalin and Nikita Khrushchev endorsed his theories. Today, however, Lysenko's ideas are so discredited that he is blamed for slowing the Soviet Union's growth in agriculture and retarding Russian research in genetics for twenty years.

The main aspect of Lysenko's theories was that species could inherit acquired characteristics. This meant, for example, that if one kept cutting off the tails of mice, eventually mice born from such a group would be tailless. Thus, according to Lysenko, new species could be created from old species by altering the species' environment; also, organisms could be trained to change. He rejected the concept widely accepted by other biologists that genes determine heredity.

Stalin liked Lysenko's ideas because they promised quick results and because they seemed to parallel Communist thinking in other areas. Lysenko believed that just as in the Soviet Union there was no struggle among the classes, so, too, no struggle existed within a species.

In 1948, for example, he advocated that a way to plant forest belts in the USSR's dry interior was to plant trees in clusters; the trees would be better able to protect themselves and each other against other species. But the dry climate doomed the trees, and they died.

Lysenko's reputation had been built on his experiments in 1929 with winter wheat seeds. He sought to demonstrate that these seeds could grow in the spring if they were soaked prior to planting. The growing patterns of other seeds could also be altered in this way.

This concept was in accord with Lysenko's idea that the conditions under which a seed is cultivated, not its genetic composition, determine the length of time it takes for a seed to germinate and grow. If true, this would have been a boon to Soviet agriculture, speeding up the time it took to plant and harvest crops in the harsh climate.

The problem was that Lysenko's experiments were not conducted in a professional manner. In his most famous experiment, for example, he used only two plants, and one of these died.

But with his growing favor in the eyes of Stalin and as president of the V.I. Lenin All-Union Academy of Agricultural Sciences, Lysenko was able to still any opposition. Beginning in 1935, he criticized the science of genetics, which was based on Mendel's laws of heredity, and by 1948 he was able to outlaw even the study of genetics. Scientists who opposed him were either dismissed from their jobs or jailed. The teaching of biology in the Soviet school system was changed to agree with Lysenko's theories.

Only after the ouster of Khrushchev in 1964 were Lysenko's theories finally allowed to be discredited publicly. The study of genetics then returned to Soviet science, with a journal published on genetics and the formation of the Institute of General Genetics. But Lysenkoism had taken its toll. Soviet farm production, which Stalin and Khrushchev had sought to improve through use of Lysenko's ideas, had been damaged. And Soviet research into not only genetics but the whole field of biology had been seriously set back.

EINSTEIN'S DOUBLE ERROR

The most widely accepted explanation among scientists today about the origin of the universe is the big bang theory. According to this theory, at some time—now calculated as approaching 20 billion years ago—a great explosion of matter created the building blocks of our world, a world that, with galaxies moving rapidly away from one another, has been expanding ever since because of that first cosmic explosion.

One scientist who was late to accept the big bang theory was Albert Einstein—and he was a holdout because of not one but two errors he made.

Although his own theory of relativity predicted an expanding universe, Einstein overlooked this support for the concept. Instead, he pursued a mathematical approach to solving the problem of a universe in expansion, but in doing so he committed, according to the scientist Robert Jastrow in *God and the Astronomers* (W.W. Norton, 1978), "a schoolboy error in algebra." What Einstein did was divide by zero during his calculations, "a no-no in mathematics." When a Russian mathematician, Alexander Friedmann, pointed out the error to Einstein, the missing solution to the expanding universe "popped out."

But Einstein was hardly pleased at Friedmann's discovery of his slipup. The great scientist simply ignored Friedmann's letter pointing out the mistake, and later, when Friedmann published the results of his findings in *Zeitschrift für Physik,* Einstein wrote a follow-up letter to the scientific journal with other calculations "proving that Friedmann was wrong." Actually Einstein had erred again.

It was then that Friedmann again wrote to Einstein, this time with great politeness and pointing out as delicately as he could that Einstein had made another mistake. After correcting Einstein's algebra, Friedmann wrote, "Most honored professor, do not hesitate to let me know whether the calculations presented in this letter are correct."

But Friedmann, knowing he was right and the great Einstein was wrong, went on. "In the case that you find my calculations to be correct, you will perhaps submit a correction."

In a 1923 issue of the scientific journal in which Friedmann had first published his findings, Einstein finally acknowledged the two errors he had made. "My objection [to the first Friedmann letter] rested on an error in calculation. I consider Mr. Friedmann's results to be correct and illuminating."

Even though his own theory of relativity gave support to the big bang theory and the concept of an expanding universe, Einstein never did like the idea that the universe had started with a bang and its galaxies were now flying apart. "To admit such possibilities," he once wrote, "seems senseless."

WHAT KIND OF STUDENT WAS EINSTEIN? DON'T ASK.

How smart was Einstein as a student? Here is how three biographies answer the question, showing thereby the difficulty in preventing mistakes from entering scholarly works since all three books cannot be correct.

From *The Drama of Albert Einstein* (Doubleday, 1954) by Antonina Vallentin, who knew Einstein personally over many years: "All the members of the family who knew Albert as a child or heard his elders speak of him described him as almost backward. . . . [He] revealed a deliberate slowness in all that he did, which irritated his teachers." Later, after he had become interested in Euclidean geometry because of the gift of a book, he "made great progress in mathematics at school but was backward in most of the other subjects. There was nothing to draw attention to him. . . . His former teachers, in fact, did not even remember having had him in their classes."

From *Einstein: The Life and Times* (World Publishing, 1971), a 718-page biography by Ronald W. Clark: ". . . Einstein's son Hans Albert . . . says his father was withdrawn from the world even as a boy—a pupil for whom teachers held out only poor prospects." According to Clark, Hans related the story that a teacher, asked by Albert's father about the boy's future, remarked, "It doesn't matter what he does, he will never amount to anything." Clark further writes: "Einstein became, as far as the professional staff of the [Swiss Federal Polytechnic School] was concerned, one of the awkward scholars who might or might not graduate but who in either case was a great deal of trouble."

From *"Subtle Is the Lord . . .": The Science and the Life of Albert Einstein* (Oxford University Press, 1982) by Abraham Pais, who, himself an eminent physicist, worked alongside Einstein in the years after World War II: "At about age six he entered public school. . . . He did very well. In August 1886, Pauline wrote to her mother: 'Yesterday Albert received his grades, he was again number one, his report card was brilliant.' . . . The infant who at first was slow to speak . . . becomes number one at school (the widespread belief that he was a poor pupil is unfounded). . . ."*

KAPTEYN OF THE UNIVERSE?

Centuries ago many people believed that the earth was the center of the universe. Today we know this is not true, but for fifteen years during the twentieth century, many scientists believed that our solar system was the center of the entire universe.

The reasons for this belief were the observations of Jacobus Kap-

---

*The dispute about Einstein's scholastic abilities as a young student appears to have been resolved by the acquisition in 1984 of Einstein's academic records from Swiss archives. As reported in the *New York Times* of February 14, 1984, the records "confirm that Einstein was a child prodigy, conversant in college physics before he was 11 years old, a 'brilliant' violin player who got high marks in Latin and Greek." His one problem area was French, which could have caused him to fail his college entrance examinations.

teyn, one of the pioneer explorers of our galaxy. However, he made a mistake that led to the fifteen-year debate over whether our universe is sun-centered or not.

In an effort to determine the position of our galaxy in the universe, Kapteyn counted the number of stars visible in various directions in the sky. On the basis of these observations, he concluded that our solar system was in the center of the galaxy and that it might therefore be in the center of the universe. This concept came to be known as the Kapteyn universe.

Although many astronomers accepted this view, others were uneasy about the idea—and among them, to his credit, was Kapteyn. His uneasiness was for good reason.

Vesto Melvin Slipher, the scientist who first discovered that the galaxies were speeding away from one another, eventually discovered Kapteyn's error. Our galaxy appears to be at the center of the universe only because thick clouds of dust out in space prevent our seeing the light from the many stars near the real center of the galaxy.

However, it took fifteen years for Slipher's findings to end the debate over the Kapteyn universe.

THE GUINEA PIG ISN'T

The term *guinea pig* is often used to mean "experimental animal," but the truth is that at least four other animals are used more often for scientific and medical experiments.

The leading experimental animal is the mouse, followed by the rat, the chick, and the rabbit. The guinea pig ranks only fifth.

Besides, this animal's name is a mistake, too. It's not from Guinea, and it's not a pig. Guinea pigs, which come from South America, are actually part of the rodent family.

MINUS A MINUS EQUALS MILLIONS

The crowning achievement of science in the modern age may well be the space program. After centuries of longing to reach the skies, humanity is no longer bound to this world. Science has freed it to fly, to soar, to speed through space.

Except that science has not freed itself of the glitches that can affect the best-laid plans of scientists.

The *Mariner I* space probe is a case in point.

Launched on July 22, 1962, from Cape Canaveral, Florida, *Mariner I* was programmed to provide the first close-up view of the planet Venus. The flight plan was a thing of beauty. Thirteen minutes into the launch a booster engine would accelerate the spaceship to 25,820 miles per hour. Forty-four minutes into the trip 9,800 solar cells would unfold. Eighty days into the flight a computer would calculate the final course corrections. And after 100 days the craft would circle the mysterious Venus, scanning the heavy clouds which enwrap it, trying to peer at the surface below.

That was the plan.

What actually happened was somewhat less successful. *Mariner I* plunged into the Atlantic Ocean—four minutes after takeoff.

An investigation later revealed that a minus sign had been omitted by mistake from the instructions fed into the computer.

A launch spokesman explained how an ultramodern computer had given an ultramodern spaceship the wrong instructions. "It was," he said, "human error."

It was also an $18.5 million loss for the U.S. space program.

"FLOP OF THE CENTURY!"

The astronomers were predicting a grand and glorious sight. In January 1974, they said, one of the most dramatic comets ever seen would fly by the earth. Harvard astronomer Fred Whipple predicted that Comet Kohoutek, as it was called, "may well be the comet of the century."

An excited public began buying up telescopes and binoculars for the big day. Even Kohoutek T-shirts were printed up as part of the festive anticipation.

After all, astronomers were saying that Kohoutek had a tail 50 million miles long, that as it flew by, this tail alone would stretch one-sixth of the way across the sky, and that the head of the comet would glow five times brighter than the moon. In January, when the comet came closest to the sun, Kohoutek would be the brightest thing in the heavens.

So what happened?

Nothing.

At the appointed time people scanned the sky, but most found no trace of a comet. Only those with binoculars could see a faint smudge near Venus and Jupiter, which appeared, as usual, as the brightest objects in the sky. Certainly no 50-million-mile-long tail blazed by.

Although various scientific reasons were offered for the faintness of Kohoutek, none proved adequate for explaining the disparity between astronomers' enthusiastic prediction and the comet's woeful performance. Later *Time* magazine characterized the comet as "a disappointing dud" and quoted a band of amateur astronomers, gathered in frigid weather atop the Empire State Building to watch for the comet, who told a TV interviewer their feelings about Kohoutek. It was, they agreed in unison, "the flop of the century!"

FORECASTING THE WINTER OF THE CENTURY

The winter of 1982–83, according to a chorus of weather experts, was to be the coldest winter of the century. Hurd C. Willett, meteorologist at the Massachusetts Institute of Technology, who has a strong following in the forecasting world, was telling everyone in September 1982 that, as one reporter later wrote, "the winter of 1982–83 was going to make the previous winter seem like moonlight in Miami."

And once Willett had offered his prediction, other meteorologists came forward, offering corroborating evidence. Excessive amounts of volcanic ash, most notably from El Chichón in Mexico, were noticed in the atmosphere, and this, it was said, would cut off much sunlight, thereby adding to the cooling down of the weather-to-be. Almanac-oriented weather watchers noted that bears and beavers had thicker coats, said to be a preparation for colder times ahead.

What did happen?

The 1982–83 winter turned out to be, especially on the East Coast, one of the warmest of the century. Except for one major storm at the end of February (and even that at the time was a surprise to forecasters), temperatures ran above normal and precipitation was below average.

One of the few weather forecasters that had not gone along with the predictions of a disastrous winter was the U.S. Weather Service, which had cautioned against relying so heavily on such a long-term forecast.

"We feel that you really can't make an accurate prediction that much in advance," said Fred Davis, head of the Weather Service's office at Baltimore-Washington International Airport in a Baltimore *Sun* interview as the mild winter wound down. "Our long range forecasts are for 90 days, and in November we predicted pretty much what happened."

The Weather Service itself had recently found that the surface water temperatures of the Pacific Ocean often give an accurate forecast of the weather that moves from west to east across the United States. The service has therefore turned to checking these water temperatures before making forecasts.

"If Willett and the others had done that, they would have realized that there just wasn't going to be that much cold weather in the east this winter," the reporter noted.

The headline of the article reporting this forecaster flub was SO WHOEVER SAID EXPERTS WERE RIGHT?

Discoveries Made by Mistake

Mistakes abound in science, but many mistakes have led to boundless good—in the form of much-needed inventions or discoveries. Here are some:

WHEN GOODYEAR'S WIFE ALMOST SAID GOODRIDDANCE

Charles Goodyear had been working for years, and sacrificing his family's finances and his own health, in his search for a way in which rubber could be made heat-resistant and thereby a practical material. Although promising his long-suffering wife, Clarissa, that he would either stop tinkering with rubber or get a job to feed the family while working on the problem in his free time, Goodyear one day began experimenting with yet another batch while his wife was out shopping. He was mixing rubber with sulfur when he suddenly heard her returning home.

To hide his work, he quickly gathered up the mixture and threw it into the kitchen stove, which was hot. Clarissa came into the house, went to the kitchen, stayed awhile, and then left to go elsewhere in the house. Goodyear hurried to the stove, opened it, and—to his surprise —discovered the solution to the problem he had long sought. By accident, Goodyear had found that rubber mixed with sulfur and subjected to great heat would turn into vulcanized rubber—the heat-resistant material he had long sought.

THE DISCOVERY THAT RANG A BELL

Alexander Graham Bell patented his telephone in 1876, but this was actually an improvement of a device he happened across in 1875. On June 2, 1875, he accidentally discovered that a telegraph transmitter using harmonic frequencies in matching receivers could transmit matching sounds. Thus was born, inadvertently, the idea that eventually gave birth to the telephone.

TWO MEN WITH X-RAY VISION

Wilhelm Roentgen, a Bavarian physicist, discovered X rays by accident. On November 8, 1895, a cathode-ray tube he had left on a barium-plated paper began to fluoresce. This continued even when he put black paper around the tube.

The next year Henri Becquerel, a French physicist, was engaged in research to find out if rays like X rays were emitted by "fluorescent bodies under the action of light." One day he mistakenly left in a drawer some uranium ore on a sensitized plate. When he came back, he found the plate had been affected, even in the absence of visible light. What he had inadvertently discovered was that uranium gave off rays with the same, but weaker, penetrating power as X rays. Marie Curie later named this discovery radioactivity.

In 1903 Becquerel, along with the Curies, won the Nobel Prize in physics for their combined work and discoveries in the field of radioactivity.

HOW SWEET IT WAS

Saccharin, a chemical substance used as a sugar substitute by diabetics or dieters because it has 400 times the sweetness of sugar but no calories, was discovered in 1879 by accident. After a day at his laboratory in the Johns Hopkins University where he was working on an oxidation experiment, Constantin Fahlberg, a chemist, washed up and went home to dinner. As he was eating, he noticed the bread tasted sweet, then examined his hands and found the sweetness not only on his hands but even on his arms. Realizing the taste was coming from something he had worked on, he went back to the laboratory and tested each beaker and basin in his work area until he found the source of the sweetness. The saccharin used commercially is basically the substance Fahlberg happened across that day in his laboratory.

COOKING WITH WAVES

Microwave cooking was discovered by accident. Percy Le Baron Spencer, of the Raytheon Company, had a candy bar in his pocket when he realized that microwave signals had melted the candy. Raytheon later used Spencer's experience to create a microwave oven which was first marketed in 1947.

NOT SO SILLY PUTTY

Silly Putty, the material that is used in toys as well as in the U.S. space program, was discovered by mistake. During World War II laboratory scientists were trying to develop a synthetic rubber when some boric acid was accidentally dropped into silicone oil, thereby creating the strange substance. Silly Putty became an enormously successful toy, but it also has proved very practical, especially in the space program. The *Apollo 8* astronauts, carrying the putty into space on their moon voyage, used it to fasten down tools during weightlessness.

THE MIRACLE DRUG THAT FLEW IN THE WINDOW

In 1928 a British bacteriologist at St. Mary's Hospital in London was conducting experiments with bacteria cultures. He accidentally left them by an open window one day, and when he later returned to his laboratory, he noticed that bits of fungus had flown into the room through the window and had landed on the bacterial culture. When he looked further, he discovered that no bacteria had grown around these bits of fungus. From this chance beginning, Alexander (later Sir Alexander) Fleming soon realized he had stumbled across a possible miracle drug to combat disease. Penicillin, as the drug was called, became so widely used that by 1950 it was being prescribed for 60 percent of all patients in the United States.

Science is, ultimately, humanity's attempt to alter nature. It may be salutary for the preservation of the human being's humility to point out, however, that . . .

You Can't Always Fool with Mother Nature

KUDZU—THE MISTAKEN MIRACLE

A half century ago an Asian vine called kudzu, with broad leaves, small purple flowers, and a thick stem, was planted in the South as a means of improving the erosion-torn soil. The perennial legume was seen as a "miracle vine" for the way in which its long tendrils and deep roots gripped the soil and replaced nitrogen in barren fields.

Hailed as the savior of the South, kudzu was promoted widely. The Soil Conservation Service gave it away free to farmers, and communities engaged in kudzu planting projects.

Today, however, the kudzu vine is an embarrassment—and a threat. The planting projects have stopped. Kudzu clubs have been abolished.

Why? Because kudzu won't stop growing. In one summer a single kudzu stem can grow sixty feet in every direction and cover the ground four feet deep. And every spring the growing starts over again.

Under ideal conditions, kudzu vines grow up to a foot a day, creeping over anything in their path—trees, utility poles, gardens, buildings, old cars.

To stop the vine, people have tried hatchets, chain saws, fire, and various chemicals—to no avail.

By now kudzu lies across 7 million acres in twelve southern states.

Although kudzu has been found to be high in vitamin A and the Japanese value it as a producer of fiber, people who have lived with the vine are leery of it in any form.

A chemical has been found to combat kudzu. Tordon 10K, produced by Dow Chemical, kills the roots if applied in the spring and fall. But the vine's wide growth over fifty years, its hardiness and its ever-continuing proliferation portend that the plant—once thought to be a miracle but now called by poet James Dickey "a vegetal form of cancer" —will never be completely eradicated but will continue to grow.

ASWAN—A DAM SITE MADE WORSE

It was intended to be one of the great feats of modern science and engineering. It would bring more fertile land and higher income to a country and a people desperately in need of both.

These were just some of the benefits envisioned by the construction of the Aswan High Dam—a feat that took fifteen years, involved the relocation of 90,000 Egyptians and Sudanese in the Nile Valley to land 400 miles north of the dam, required the moving of massive sculptures at Abu Simbel, and eventually consumed $1.5 billion.

After all this had been done and the dam had been finished, however, a host of unforeseen problems emerged when the dam went into operation.

The lake formed above the dam caused the Nile to spread out so far that substantial amounts of water were now being lost to evaporation rather than being available for power generation and fertilization of the valley below. Only four of the ten turbines were being powered, and only half the land was being irrigated.

The water that was available had a higher concentration of salt unsuitable for some crops, forcing many farmers to leave their lands.

A vital industry was also destroyed. Sardines from the Mediterranean Sea now no longer entered the mouth of the Nile.

Also, the dam was found to be promoting disease. Schistosomiasis, spread through blood flukes, began reaching an epidemic level. The blood flukes, enjoying the wet fields created by the dam, were now able to travel to all the newly irrigated lands.

But the most ironic effect of the Aswan High Dam was what it did to the land previously fertilized each year when the Nile flooded its banks. Now Egyptian farmers had to buy chemical fertilizers to do what the Nile had done naturally for thousands of years.

In other words, science was being called in to solve the mistakes of science.

"PEOPLE WHO HAVE BECOME FAMOUS THROUGH ERROR" DEPARTMENT:

THE STORY OF WRONG WAY CORRIGAN

Sometimes people have become famous only because of their errors. Take the case of Douglas Corrigan, better known as Wrong Way Corrigan.

In 1938 Corrigan was an airplane mechanic who idolized Charles Lindbergh and wanted to emulate his solo crossing of the Atlantic. But other people wanted to do the same, and authorities were skittish about permitting such flights because of their inherent danger and because Amelia Earhart, the aviatrix, had disappeared just the year before during a solo flight over the Pacific.

After air traffic controllers had turned down his request for permission to fly across the Atlantic, Corrigan announced that instead he would fly solo nonstop from New York to California.

In the early morning of July 16, 1938, the airplane mechanic took off in a plane with a leaky gas tank and headed west. Foggy conditions made visibility poor and obscured the land below.

When Corrigan went to check his compass, he found that it was not operating properly, so he consulted a backup compass he had brought aboard. It was then that according to Corrigan, he made his error. He followed the wrong end of the needle. Instead of flying westward, he began flying eastward— out over the Atlantic Ocean.

For the rest of that day and into the night he continued to fly eastward. Later he said that he thought the reason he did not see land was because of the fog and heavy cloud cover. Not until twenty-six hours had passed did Corrigan discover his error. He finally put his plane down in Ireland twenty-seven hours after taking off from New York for a flight which was supposed to end on the west coast of the United States but wound up on virtually the west coast of Europe.

At first few people believed Corrigan's story that he had committed a twenty-six-hour navigational mistake. Air traffic controllers in New York who knew of Corrigan's desire to emulate Lindbergh had seen the flier turn his plane eastward in violation of his flight plan, and they had alerted authorities in Ireland, who now held him for questioning.

Corrigan, however, stuck by his story, even passing various lie detector tests. In the public debate that arose over whether or not Corrigan was telling the truth about his error, it was noted in his defense that he had not prepared himself for a transatlantic flight. Unlike Lindbergh, who had carefully figured out food and clothing needs, taken maps, and prepared his plane especially for the long journey, Corrigan had not taken any water or enough food, had no warm clothing, and had not even carried maps of the ocean and its coasts.

Wrong Way Corrigan may very well have been aptly named. If so, his achievement was even more impressive. He did by careless mistake what Lindbergh could do only after careful planning. Such is error's hidden power for good . . . sometimes.

CHAPTER IV

Snafu:
Error in Government

*The bad news is that our government is
predicting a recession. The good news is our
government hasn't been right yet.*

—Johnny Carson,
television talk show host,
on *The Tonight Show*

The bored government clerk watching the clock. The underpaid bureaucrat sitting quietly behind his gray metal desk. The listless
laborer leaning on his shovel. The elected official concerned primarily
with reelection.

Add to this abundant amounts of red tape that trip the unwary, an
oversupply of poorly planned pork-barrel projects, an ample number of
convoluted laws.

These are some of the images many of us have about government
—local, state, or federal. To the average citizen, government appears
to be a breeding ground for waste, inefficiency—and foul-up.

How much truth is there to this conception?

Government is no better than the people who staff it. If they are
bored, befuddled, underpaid, or overworked, their product—the ser-

vices of government—suffers. In government the vicious cycle picks up speed. Noncaring leads to carelessness; erratic work leads to error-filled results.

There is the famous incident from the nineteenth century of the congressional clerk who made a multimillion-dollar error in the transcription of a bill. "All foreign fruit-plants are free from duty" was the way the newly passed law read, but the clerk accidentally changed the hyphen to a comma and wrote, "All foreign fruit, plants are free from duty." When the mistake was discovered, a new law had to be passed to correct the error, but by the time Congress was able to pass the bill the United States government had lost more than $2 million in taxes.

Over the years mistakes have grown more costly for government. In 1983 the General Accounting Office (GAO), the investigative arm of Congress, discovered that the armed services and the Defense Logistics Agency, which handles supplies for the military, had been hiding billions of dollars in inventory errors. The GAO cited instances in which tanks, missiles, aircraft, and ships were out of service because they lacked parts that inaccurate records had indicated were available. At the same time that needed items were not available, the armed services were purchasing millions of dollars of unneeded items. Fraud or criminal activity was ruled out. Ruled in were billions of dollars in snafus. The chairman of the House Armed Services Committee's Readiness Subcommittee called the report "shocking" and declared that "it clearly strikes at the management of the defense supply system."

Even a part of government—the educational system—that should know better can't get its facts and figures right. Consider the time when the New York City Board of Education erred in counting the number of handicapped students it serves—and thereby stood to lose $30 million in state education aid for the 1982–83 school year. The state legislature, in enacting its education-aid formula, had relied on estimates on the number of handicapped students enrolled, made the previous year by the city Board of Education, but the estimates were never corrected with the actual—and higher—totals. Even though the state had written a complicated aid formula to give New York City a $112 million increase in education aid, the formula wound up giving the city only $82 million. Asked how it had happened that the city stood to lose $30 million, a member on the mayor's staff told the *New York Times,* "Because the Board of Education can't count."

Thus does error creep into the governmental process, affecting the citizen and frustrating the taxpayer, contributing to waste and adding to costs. In the halls of Congress, in the corridors of City Hall, in the assembly rooms of the Statehouse, the kingdom of error has full reign.

WILL TAXES BE THE DEATH OF US?

According to the cliché, we can be assured that two things will always be with us: death and taxes. But the way the Internal Revenue Service (IRS) runs things, we may yet find that taxes will be the death of us.

The IRS is not the largest federal agency, but it is the one that, no matter what our status or state, touches each of us every year and handles that most sensitive part of ourselves—our money.

It may therefore be surprising to learn that the IRS is as error-prone as taxpayers are.

Thanks to a Washington couple by the name of Philip and Susan Long, we now know that the Internal Revenue Service makes innumerable errors, especially when giving advice to taxpayers. The Longs conducted a decade-long campaign to force the IRS to make public many of its previously secret documents. Using the Freedom of Information Act, as of mid-1981 the Longs proved successful in their disclosure suits twelve out of twelve times, with a thirteenth attempt pending. According to the Longs, this effort uncovered two major findings, neither of much consolation to us taxpayers:

1. Professional income tax preparers frequently make mistakes.
2. The IRS also makes mistakes, especially when it gives tax advice, in which the IRS is as error-prone as tax return preparers.

Reader's Digest, in a February 1979 article entitled "IRS Tax Advice: It Ain't Necessarily Sound," warned its readers that "caution is in order for all who ask the IRS for help in preparing income tax returns" and pointed out that the agency "makes a disturbing number" of mistakes.

The *Digest* cited a project conducted by the *Wall Street Journal* in which various IRS offices were consulted for advice on the same income tax return. While one office said the government owed a refund of $812, another said the taxpayer owed the government an additional tax of $52. In all, IRS personnel at five different locations gave five different replies. They could not even agree on how many different forms had to be filled out.

In another study a private group asked IRS offices in six states and the District of Columbia for help in filling out an itemized joint return for a married couple with one child. The various IRS personnel gave answers varying by as much as $300 in either taxes due or to be refunded.

Reader's Digest cited IRS practices involving misleading or incorrect advice in its own publications (noted was a 1977 publication, *Your Federal Income Tax,* with 2.7 million copies distributed, which had faulty advice on deducting property losses) and the use of poorly trained personnel employed to give information over the phone (in one study conducted by the General Accounting Office, in which agents posed as puzzled taxpayers, an error rate of 13 percent was discovered in answers provided by IRS staffs).

The *Digest* concluded, "The shocking truth is that agency employees can and do make mathematical errors, distribute the wrong forms, offer incorrect advice, overstep their bounds and even change the rules retroactively."

Taxpayers, however, face one problem in dealing with IRS errors. The agency is not responsible for the mistakes it makes. A taxpayer relying on erroneous IRS advice cannot use it as a defense in an audit.

And the unkindest cut of all is that as the *Digest* noted, the IRS's mistakes in providing erroneous information "increase taxes unfairly for some, while it lets others pay far less than they owe."

WHY SOCIAL SECURITY MAY NOT REACH ITS OWN SIXTY-FIFTH BIRTHDAY

The United States Social Security system has been in deep fiscal trouble in recent years. It seems that people are living longer than the retirement age of sixty-five, thereby draining the benefit pool. Studies are under way on how best to deal with this problem, but few people seem to realize that the problem emerged out of a basic mistake—the mistake of even choosing age sixty-five.

Where did the choice of sixty-five come from?

Director, producer, and author Garson Kanin was moved to find out how the government arrived at this figure. In an interview with columnist Phyllis Battelle in 1977, he told what he had discovered and was writing about in his forthcoming book *It Takes a Long Time to Become Young* (Doubleday, 1978).

"I went to Washington and asked senators and congressmen: 'Why 65?' Nobody knew. I researched the libraries. No help. Finally I tracked down one of the original writers of the Social Security Act. . . ."

What Kanin found was that the idea of using sixty-five as a beginning age for drawing Social Security benefits was first started by German Chancellor Otto von Bismarck in the late nineteenth century. In 1881 a crisis was gripping Germany and socialism was spreading among German workers. Bismarck met with his advisors to determine a way to head off the growth of socialism. It was noted that among the socialists' programs was the promise of old age benefits for the people. Bismarck's response was to arrange for the government to provide workers with a pension, up to one-half their salaries, at a certain age. But then the question became: What age?

"They decided that 65 sounded like a good time," Kanin said, "because what the chancellor and the brain trusts knew that the people didn't know is that very few German workers lived to be age 65."

Bismarck's purpose in founding the first known social security program was therefore not so much to offer a sense of security for the German worker as it was to stave off socialism. And the age of sixty-five had been selected precisely because actuaries had told him that few Germans then lived much beyond that age. Indeed, it has been estimated, according to Kanin, that today's corresponding figure would be a hundred and five.

But times—and nutrition—changed, and so did life expectancy. The age of sixty-five, right for Bismarck's purposes, may not have been right for the U.S. government's purposes fifty years later.

Is it any wonder, then, why Social Security in the United States is today, 100 years after Bismarck, in trouble?

THE SEAT OF U.S. GOVERNMENT FACES THE WRONG WAY

When Pierre L'Enfant was designing Washington, D.C., and trying to figure out where to put the Capitol building, he decided to face it to the east. He expected that the new city would grow first in an easterly direction around the commercially promising Anacostia Harbor. He was proved wrong. The city developed not eastward but westward. Capitol Hill therefore faces away from the major developed areas of the city.

NOT MUCH INTELLIGENCE

How good is the United States intelligence service? George Keegan should know. For thirty-three years he was deeply involved in America's intelligence operation. His specialty was monitoring the Russian arms buildup. In fact, he rose high enough to have become Major General George J. Keegan, Jr., chief of Air Force Intelligence.

So what was his assessment after he had stepped down from this vital post? In an interview with reporters Stanley A. Blumberg and Gwinn Owens, cited in an article published in the Baltimore *Sun* in early 1983, General Keegan declared that U.S. intelligence judgments "were all wrong—every single one of them.

"They were wrong just as Vannevar Bush [American engineer and president of Carnegie Institution who pioneered in computer technology] was wrong in saying the ICBM would never fly, and just as many others concluded a thermonuclear weapon was not possible. . . .

"This has been the history of American intelligence. In my 33 years of direct involvement in the intelligence business, we have been wrong in every single major scientific prediction."

FILED AND FORGOTTEN

During World War II, with their supplies of oil cut off, the Germans, after years of effort, developed a way to derive gasoline from coal. In fact, from 1944 until the end of the war Nazi tanks ran almost exclusively on coal-derived gasoline.

In 1975, at the height of the oil crisis, a chemical engineer at Texas A&M, along with a colleague, decided to look at captured German documents to see what they could learn about the Germans' synthetic oil program. The two found that the Nazis' voluminous records had been filed and forgotten. Some were at the National Archives in Washington. Others were stuffed into crates in government buildings around the country. Even though the war and subsequent events in the Mideast showed the importance of synthetic fuel technology, until the two professors came along, the documents had lain untouched in government files for thirty years.

THE $1.6 BILLION ERROR

During the 1960s Great Britain suffered from dramatically declining trade deficit figures which made its economy plummet and its standard of living dwindle. But then, in 1969, the British government discovered an error in its accounting procedures. The government had actually been cheating itself out of $24 million a month—for six straight years. It was a $1.6 billion error. Roy Jenkins, chancellor of the exchequer, announced that the government had been erroneously underestimating the value of British exports by 2 to 3 percent over the previous half dozen years. Instead of being deficit-ridden, as beleaguered Britons had been led to believe, the country was almost certainly in the black during the first five months of 1969, said Jenkins. And for the calendar year 1968, Britain's overall balance of payments deficit was nowhere near the $1.3 billion originally forecast. Thus, many of the British government's austerity measures placed upon its citizens had been unnecessary.

MOST ASININE MISSTATEMENT BY A GOVERNMENT OFFICIAL

By the 1950s the people of America were in a state of near panic every summer. Polio was becoming the scourge of children and their parents. More than 50,000 cases were being reported each year, with hundreds of deaths annually and many young people doomed to spend years in iron lungs or walking with braces. Then, in 1954, a vaccine discovered by Dr. Jonas Salk to guard against polio was successfully tested on one million people. As a result, the next year the U.S. government began to offer the vaccine on a wide scale and an eagerly waiting citizenry flocked for inoculations. Soon, however, the government ran out of the vaccine.

Oveta Culp Hobby was then the secretary of health, education and welfare, serving in the administration of President Dwight David Eisenhower. When asked why the government's supply of the vaccine had been so quickly depleted, Mrs. Hobby replied, "No one could have foreseen the public demand for the Salk vaccine."

OUR FAVORITE POST OFFICE STORY

Among the agencies of the government the postal system has always come in for a great share of horror stories about inefficiency, especially tales about long-overdue deliveries of mail. One of the best stories in this regard involved a soldier in Vietnam, George Mellendorf, who sent a letter on January 4, 1971, to then President Richard Nixon complaining about the slow delivery of mail to the soldiers. Wrote Mellendorf:

"It seems nobody cares if we get our mail. We are lucky to get it twice a week. Sir, someone is not doing their job."

Mellendorf's letter was finally delivered to President Nixon in February 1978, more than seven years after the soldier had sent his missive criticizing postal delays.

LET THEM EAT CASSEROLE

One letter writer to a president got through soon enough but received the wrong letter in reply. In 1981 an Illinois woman wrote to President Ronald Reagan to complain about the unemployment situation. In response she received from the White House a form reply—but something went awry, and instead of a message on unemployment, she got the Reagans' favorite recipe for crab meat casserole.

According to a *Newsweek* magazine item ("White House Pen Pal," March 22, 1982), the bungled letter caused the president "considerable embarrassment." After the foul-up had been discovered, the correspondence unit of the White House, which organizes incoming mail for appropriate presidential responses, decided to head off any similar embarrassment by taking the following move: It substituted for the crab meat, said *Newsweek,* "a more plebeian recipe for macaroni and cheese."

YOUR GOVERNMENT INSPECTOR

The mistakes made at the Three Mile Island nuclear power plant—which have caused the loss of billions of dollars—are well known to us all. But less well known are the mistakes that were subsequently made at the Nine Mile 2 nuclear power plant. On November 2, 1982, the federal Nuclear Regulatory Commission found this plant near Oswego, New York, to have more than 1,300 welds that were improperly inspected. Officials said the problem was caused by the use of trainees rather than certified inspectors to make the inspections. As a result, 1,383 welds had to be reinspected.

SISTER, CAN YOU SPARE A DOLLAR? THE SUSAN B. ANTHONY FIASCO

In January 1979 the U.S. Bureau of the Mint began producing the Susan B. Anthony dollar, a metal coin about the same size as a quarter. The mint struck 857 million coins, each with the famous suffragist's picture on the front side and the *Apollo 11* moon landing on the reverse.

According to a Federal Reserve spokesman, the coin would save taxpayers an estimated $50 million a year because fewer paper dollars would have to be produced and the coin would last longer.

But two market studies had warned the Treasury that success of the new coin was doubtful. Indeed, the Susan B. Anthony dollar never caught on with the public, in large part because of its similarity in size to the quarter. It quickly became, in the words of the *Washington Post,* "a federal white elephant." On September 26, 1979, just nine months after the mint had begun issuing the coin, Jack Anderson in his nationally syndicated column was calling it "a costly fiasco, rejected by merchants, bankers, and the public alike."

By March 1980 production of the Anthony dollar had been stopped. At the time the government still had more than 500 million of them on hand.

In 1982 the government admitted that the Anthony dollar was "a disappointment." Mint spokesman Frank DeLeo said, "It's obviously not working. We're not realizing the savings that were anticipated." At the time the mint still had 364 million Susan B. Anthony coins in its vaults and Congress had not appropriated funds to continue their production.

This was not the only mistake the Treasury had made in recent times trying to alter Americans' monetary habits. The government had also laid eggs with the issuance of the Eisenhower dollar, the Kennedy half-dollar and a new two-dollar bill.

GIVE US OUR DALEY QUOTE

When rioting broke out in the streets of Chicago during the 1968 Democratic convention, Mayor Richard Daley called out the police, who battled openly with rioters. In later commenting on the action of officers, Daley declared, "The policeman isn't there to create disorder. The policeman is there to preserve disorder."

WHEN GRANTS WERE MISTAKEN AND MISTAKES WERE GRANTED

On November 21, 1981, newspapers across the United States reported that nearly half the 2.7 million college students who received federal grants in 1980–81 had been paid more than they should have. The errors cost taxpayers $300 million.

U.S. Education Secretary T. H. Bell, who made the announcment, said he based his estimate of the $300 million in overpayments on a study of mistakes in federally funded Pell grants to 4,000 students that year.

How had the multimillion-dollar mistake been made?

The Department of Education had been accepting on faith the claims by students about family income and other data. No system had been set up for checking the claims.

How would the multimillion-dollar mistake be prevented in the future?

Secretary Bell promised that from now on the government would stop accepting claims on faith and would check.

THE GOVERNMENT STUDY

One of the classic areas for government mismanagement and mistake is the government study. Ostensibly made to save money, invariably, it seems, the study winds up costing money—a lot of money—and hardly ever translates into appreciable savings in the long run. Here are two examples of snafus in government studies:

—In 1977 the *U.S. National Highway Safety Administration* contracted with a California firm to build a motorcycle with a new center of gravity powered by the front wheel and steered by the back wheel. Even though the NHSA's own initial testing indicated that such a vehicle would be impossible to steer, the agency believed that perhaps it could be designed to be usable.

Four months into the contract period the California firm reported to the NHSA that its computer simulations indicated that the motorcycle would be very difficult to control. The next month the firm again asked that it be allowed to stop work on the vehicle. The agency demanded that the motorcycle be built and tested.

Two months later the motorcycle was finished, yet as predicted, it was unridable. The longest ride that experienced motorcyclists took on it lasted 2.5 seconds. In fact, training wheels had to be added to the cycle to prevent serious injury to the riders.

The cost to U.S. taxpayers of this exercise in futility: $120,126.

—In 1976 the *U.S. Law Enforcement Assistance Administration* developed a new police car, full of technical improvements, under its Police Patrol Car System Improvement Program. Unfortunately the improvements did not work correctly or were impractical. The program was stopped before field testing was begun, but not before $2 million had already been spent.

The "improvements" included a multiple spark discharge ignition system which was "prone to failure" (LEAA report), a digital audio receiver the battery of which ran out very quickly, a periscope rearview mirror which required extensive rooftop modifications, a carbon dioxide monitor, and eight different sensors of mechanical failure. It was also discovered that occupants of the car would have little room.

Even if the improvements had worked, the median price of the cars would have been $49,078.

The planners had committed one other mistake: No one had thought to consult a working policeman about the design for a new police car.

HOW THE GOVERNMENT BUILDS

The New Senate Office Building, known also as the Dirksen Building, is considered one of the great boondoggles of all times. Errors in construction resulted in a cost overrun of almost 300 percent. Just two of the mistakes: The metal hands on the main clock in the building were so elaborate and heavy (and expensive) that they could not move and had to be replaced. And the floors, originally a very expensive brand of linoleum, were found to be too slippery. The linoleum was then covered with costly plush carpeting, but the carpeting was too thick for doors to be opened and closed, so all the doors had to be removed and planed.

BLAIR HOUSE AND THE CASE OF
THE FALLING CHANDELIER

There's something about government and buildings that do not seem to go together. The White House was so poorly constructed and maintained that President Harry Truman had to vacate the executive mansion for several years while it underwent a multimillion-dollar renovation. During that time the Trumans stayed at Blair House, a three-story row house across from the White House. Before and after serving as the president's residence, Blair House has been the U.S. government's official guest residence, where visiting heads of government stay on state visits to the United States.

Over the years Blair House also was allowed to deteriorate until in August 1982 the government needed $7 million for emergency repairs to correct "hazardous conditions" that had made the building too dangerous for overnight visitors. "The life safety of these official guests cannot be ensured," said a prospectus submitted to Congress seeking the renovation funds.

Among the problems cited at Blair House: flooding of the basement, garage, pantry, and elevator shafts; severe problems with the heating and hot-water systems; leaks from the roof and air conditioning that stained walls and damaged carpeting.

Several senators grumbled at the cost of renovation, which came to $156 a square foot. But a Senate committee eventually approved the expenditures, which included: replacement of the heating, ventilation, and air conditioning, $1.26 million; modification of the interior, $674,000; rewiring the house, $662,000; redoing corridors, elevators, and other entries and exits, $423,000; and waterproofing the basement and painting inside the house, $326,000.

All these corrections and improvements were made even more necessary by the notification to the Senate committee that Blair House had to be closed to overnight guests on June 24, 1982, because a gas boiler almost exploded.

But that wasn't the only potentially explosive situation averted by the closing of Blair House. According to Senator Robert T. Stafford (R.-Vt.), chairman of the Senate Environment and Public Works Committee, shortly after the discovery of the boiler problem a chandelier in the master bedroom fell from the ceiling and crashed onto a bed in the room.

"Fortunately for us, no one happened to be in the bed at the time," the senator said.

Fortunately, indeed. The room where the chandelier fell and the bed upon which it crashed are where visiting heads of state sleep.

GOVERNMENTAL MATHEMATICS

Since government is often the keeper of statistics by which citizens live, and sometimes die, governmental mistakes in mathematics can be disquieting, to say the least.

Consider, therefore, the U.S. Census.

The 1980 census, which cost more than $1 billion, has been slow to yield its data (in fact, as of 1983, it was a year slower in issuing its results than the census conducted in 1970). But when the Census Bureau did provide data—such as the per capita income figures—the result was a glaring error. In fact, in late 1982 the bureau had to issue revised figures for 400 of the nation's 3,100 counties after it had found that some clerks had made errors in transferring information from the 1980 census forms.

The corrections necessitated changes in revenue sharing and other federal grant programs tied closely into census data. Correcting the errors, of course, also meant further delays in reporting census data, by as much as another four months.

The mistakes occurred when clerks went over the long census forms filled out by approximately 15 million of America's 80 million households. On those questions asking about income, the clerks were supposed to round off amounts to the nearest $10 and take off the last zero before filling in circles that optical scanners could read for the computers. Said Peter Bounpane, assistant director of the Census Bureau for demographic censuses: "Unfortunately, sometimes they kept in those last zeros or they filled in the circles from the left so the amounts are too large by [factors of] 10 or 100 or 1000."

The mistakes, although they had little effect on national statistics or those for large counties and states, had a substantial impact on smaller areas. One county in Georgia, for instance, had 8.6 percent of its forms incorrectly coded.

What made the Census Bureau realize it had a mistake on its hands were the complaints of about 200 local governments across the country that their revenue sharing grants were too small because their reported incomes seemed too high.

Commented Bounpane: "When you have large numbers of people doing repetitive work, something's bound to happen."

THE "AMAZING STATISTICAL GOOF"

The Associated Press termed it "an amazing statistical goof" and noted that "the government, sole tabulator of most national economic figures, just isn't supposed to get the numbers wrong."

What was wrong? On November 30, 1982, the Commerce Department announced at 10:00 A.M. that the government's important Index of Leading Economic Indicators had gained 0.6 percent. The index, which forecasts economic trends for the nation, is always eagerly awaited by the business community. And that day's healthy 0.6 percent advance was very welcome news during the recessionary period.

But by 4:00 P.M. that day Commerce Department economists, somewhat abashed, had to announce that the index had risen not 0.6 percent but a more meager 0.2 percent. It seems that some of the government's economists wondered why the figures released that morning had been better than anticipated. Another department economist then went to work and soon found the mistake: A component listing the previous month's orders to factories for consumer goods had not been as good as the figure used to determine the index.

Meantime, before the error was discovered and announced, the Dow Jones industrial average rose 36.43 points—at the time the fourth biggest one-day advance in Wall Street history.

When I make a mistake, it's a beaut.

—Fiorello LaGuardia,
mayor of New York,
about a bad judicial appointment he had made

ACTUALLY, THEY WANT THE ACTUARY TO RETIRE

The state of Maryland realized it had a potential problem on its hands. Its retirement system, unless changed, would bleed the state of funds to pay a too-lucrative pension for retired state employees. So in 1979 the Maryland retirement system was overhauled. But the reforms failed, and in 1983 the legislature had once again to wrestle with the problem. Why?

Because an actuary working on the reforms made what one columnist for the Baltimore *Sun* termed "a sixth-grade math error."

The error? The actuary's projections of the funds the new system would need in the future were based on computations using simple interest instead of compound interest. The mistake meant a difference of millions of dollars in obligations for the state of Maryland and, within just four years, a retirement system that, instead of being solvent, now faced bankruptcy.

OUR FAVORITE SOVIET GOVERNMENT CLERK STORY

When it comes to governmental snafus, no one seems to have anything on the classic Soviet government clerk. He just seems to keep on doing his blind duty, stamping away at documents and stomping away on people's rights. The result is often inefficiency at best and outright error at worst.

In this regard a clerk at a Moscow airport, no doubt just doing his or her duty, in December 1981 made a mistake that sent a Soviet couple on a 3,000-mile error.

Near the end of 1981 a Mr. and Mrs. M. Lewenetz took a flight from Moscow to the United States to visit his father. When they emerged from their airplane at the end of the trip, they expected to find warm, sunny Florida. Instead, they found themselves in frigid Alaska.

The couple did not realize their mistake, however, until they tried to get a cabdriver to take them to the address of Mr. Lewenetz's father. Later a resident of the town who spoke Russian helped straighten out the problem. The Lewenetzes arrived finally in Florida on Christmas Eve.

An official at Alaska Airlines eventually explained what had happened. The couple wound up in Alaska rather than in Florida because an airport official in Moscow had left the "St." off "St. Petersburg, Florida," in filling out the tickets, thereby sending the couple to Petersburg, Alaska, a fishing town 100 miles from Juneau.

It also figures that in atheistic Russia a government clerk would drop any reference to a saint.

SUNGLASSES AND SOAP BUBBLES—SOVIET STYLE

The Soviet Union, characteristically, does not like to admit its mistakes, so reports of missteps and mishaps in the all's-right-with-communism world are hard to come by. But one item that did leak out told about a Soviet factory, in the Ukrainian city of Donetsk, where some unusual products were manufactured. One was a pair of sunglasses. Thirteen thousand of them were produced before a defect was discovered: They were so dark that not only did they shield wearers' eyes from the sun, but the sun could not be seen even if wearers looked directly at it.

Then there were the 3,000 plastic footballs that did a curious thing when they were kicked: They burst, said the news report, like soap bubbles.

According to TASS, the Soviet news agency, the manager of this wondrous factory had been fined a month's salary.

REAL MACHINE POLITICS

The electoral process is fundamental to the operation of good government. Protecting and preserving that process are paramount obligations of democratic government. It is all the more disturbing, then, when errors occur on election day.

One glaring election day error occurred in September 1982 when the Baltimore Board of Elections discovered a 24,000-vote error in the city police department's count of the 96,000 votes cast in the race for register of wills. The mistake, which was not discovered until three days after the election, meant that the incumbent, who thought he had won reelection, had been defeated overwhelmingly by a female newcomer to politics.

Police attributed the error to a temporary employee's unfamiliarity with the type of calculating machine being used. As a result, a 24,000-vote subtotal was incorrectly counted twice for the incumbent.

A police spokesman, in reviewing the incident, said that the mistake occurred when a woman handling one of the machines used to tabulate the vote totals from the different precinct sheets "hit a button on the machine that she thought was supposed to clear it. On other [calculating] machines, the button to clear it would have been located where she hit, but not on this one. So instead of clearing it, she started a new total by adding the 24,000 votes in again."

The incumbent register of wills, it was said, was somewhat ill after learning from the Board of Elections by telephone of his defeat—three days after he had celebrated what he had been told by the board was a clear-cut victory.

"A VOTE FOR BOSTON CURTIS IS A VOTE FOR . . ."

If government can be careless, so, too, can the voters. This, anyway, was the theory of a mayor of a town in the state of Washington, who tested his idea—and was proved right in a most unusual way.

In 1938, the Democratic mayor of Milton, Washington, put a mule named Boston Curtis on the ballot in the race for Republican precinct committeeman. The candidate's hoofprints were imprinted on the filing notice, and the mayor signed as legal witness.

On the day of the election the mule won by a fifty-one vote plurality.

After the race the mayor said he had sponsored the mule in order to prove that many voters are careless in how they vote.

In this case the election showed that the candidate was not the only jackass involved.

"THE MOST MODERN JAIL IN THE UNITED STATES"

It was billed as "the most modern jail in the United States" when the $11.5 million computer-controlled detention facility was opened in Baltimore County in January 1982. In March, however, nine inmates were able to escape by kicking out a barless glass-and-plastic window that had been touted as "unbreakable."

When the escape became public, the sheriff, Charles H. Hickey, revealed to the news media the many problems being encountered with the new prison. "Please make it quite pointed that I did not plan this building. You just have no idea of how embarrassed I am about all this," said Sheriff Hickey.

Among the problems the sheriff cited:

— Remote-control cameras that were to pan the building had to be turned off after a half hour and rested for up to two hours for cooling; otherwise the motors burned out.
— Many of the locks either would not lock or, once locked, would not open.
— A mechanical door had malfunctioned and chopped off a guard's fingertip.
— Computer malfunctions had periodically trapped people in various parts of the building.
— Vaunted solar panels (to provide "the only active solar

heating system on a detention center in the United
States") had frozen solid in the midwinter months since
the opening of the jail and had been totally useless (570
of them had cost $400,000).
—Guards were unable to see into the cells from their
guardposts, and the microphones provided to them to
talk to the inmates in their cells were largely unusable
because of the noise in the cellblock.

Sheriff Hickey had tried, however, to make some adjustments in his
jail. One of his additions to the original design: He put a lock on the
front door of the prison.

WEAPONS DEVELOPMENT:
WHY DO YOU THINK THEY CALL IT A BOMB?

One of the prime functions of government is the protection of its
citizens. This protection is provided on the international scene by the
military. It is here, in the unwieldy apparatus of the armed forces, that
governments often commit their most costly errors.

Consider the development of weaponry.

During World War II torpedoes and bombs often failed to operate
as planned. In *Miracle at Midway* (McGraw-Hill, 1982), author Gordon Prange writes, "Americans prided themselves upon their technology, yet in many instances the quality of material was little short of
miserable." B-17 pilots, in submitting their reports after raids, often
wrote of mechanical snafus. Cited in *Miracles at Midway* were such
pilot comments as these: "On takeoff . . . the exhaust pipe on my
number one engine ruptured. . . ." "Only 3 bombs dropped due to
malfunction of racks." "Did not drop bombs due to failure of inter-
phone."

Here, for instance, was the weapons development experience of just
one U.S. secretary of defense, Robert S. McNamara. During his term
in office, the military:

—Tried to build the TFX variable-geometry fighter plane,
which not only never flew but almost ended
governments in Britain and Australia, which had
placed orders for it;
—Built the B-71 bomber, which crashed while on view
before reporters and television cameras;

—Tried to build the Sheridan main battle tank at a cost of billions of dollars over more than a decade before the project was terminated;

—Constructed the AR-15 rifle, which jammed so often after the first few shots that American GIs in Vietnam opted where possible for captured Soviet rifles.

Of course, we should not be too harsh on Secretary McNamara. After all, he was the head of Ford Motor Company when the Edsel was built.

THE SHIP THAT TORPEDOED ITSELF

In 1941 the British warship *Trinidad* was sailing in Arctic waters when a passing German destroyer was spotted. Immediately the crew fired a torpedo, but they neglected to account for what the icy waters would do to its steering mechanism. Speeding through the water at forty knots, the weapon began to curve slowly in an arc. The crew of the *Trinidad* watched, horrified, as the torpedo continued to curve until it began to head back toward the path of the *Trinidad*. Within moments of being fired, the torpedo slammed into the ship that had fired it. HMS *Trinidad* was damaged to such an extent that it never saw action again during the war.

THE BIRTH OF THE SNAFU

Snafu, synonymous for disorder and confusion, is the acronym for the expression *Situation Normal: All Fouled-Up* —although some dictionaries give the more earthy "All Fucked Up" as part of the basis for the expression (see *The American Heritage Dictionary of the American Language*).

According to various sources, *snafu* was coined in the British military in either 1939 or 1940. American GIs soon picked it up and applied it freely to the common military experience of mass confusion. After the war the word became widely used in nonmilitary settings. Today, snafu captures the recurring presence of error; as *Webster's New World Dictionary* states, the word connotes "mixed up as usual."

WHEN GERMANS BOMBED GERMANS

The British are not the only ones who could torpedo themselves. On February 22, 1940, the German air force bombed the German navy. A Luftwaffe bomber, having sighted two destroyers off the coast of Borkum, strafed, bombed, and harassed the ships, causing much damage before the Nazi fliers realized that the destroyers were named the *Lebrecht Maass* and the *Max Schultz.*

IS THAT WHY THEY CALL IT
THE WILD BLUE YONDER?

But the British and Germans have nothing over the American armed forces. In April 1981, when it wasn't even wartime and the U.S. Air Force was just conducting a training mission over Florida, an error resulted in a multimillion-dollar loss. According to the plan, a pilot was to shoot down an unmanned target plane. Instead, he confused the plane with another and shot down a manned F-4 Phantom fighter jet. The burning jet's two crewmen were able to eject themselves and parachute to safety before their aircraft crashed into the Gulf of Mexico.

In an investigation of the accident, the air force blamed the mishap on an inadequate briefing of the pilot, his failure to follow procedures, and, as the pilot had indicated, the similar appearances of the target plane, which was a converted F-102A, and F-4 jets.

These errors, however, had led to the complete destruction of one of the United States' most advanced fighter planes. Cost—$3.3 million.

THE GREAT TUBING AND PIPING MISTAKE

The Pentagon, building a gigantic structure to test new jet engines for the U.S. Air Force, discovered a plumbing mistake that meant the building would have to be rebuilt. The problem? The building required a great deal of intricate tubing and piping, but it seems the people designing the tubing and the people designing the piping had never consulted with each other. When it came time to link up the two systems, they did not fit the places in the building for which they had supposedly been designed. The result? Everything had to be done over again. The cost—$138 million.

This news item served as the basis for a column by *New York Times* humor columnist Russell Baker, driving him to comment, "I don't suppose anybody can make it through a whole lifetime here in the 20th century twilight zone without an occasional suspicion that he is living in a global booby hatch."

A GENERAL LEARNS A LESSON:
THE DESTRUCTION OF JAPAN'S CYCLOTRONS

On November 30, 1945, just months after World War II had ended, newspapers in America carried a story that many found hard to believe: United States occupation forces in Japan had arbitrarily destroyed five Japanese cyclotrons. Scientific and intellectual societies and many individual citizens protested the destruction, declaring it adversely affected scientific progress. Newspapers and political figures compared the act to Germany's burning of the Louvain library in 1914 and 1940. The action was denounced as "a crime against mankind" and labeled as "stupid."

One of those agreeing with this assessment was General Leslie R. Groves, head of the Manhattan Project and an officer involved in the chain of command that led to the destruction of the Japanese cyclotrons. Writing later about it in *Now It Can Be Told* (Harper & Row, 1962), General Groves ascribed the action to "a very serious error."

The sequence of events leading up to this major faux pas in America's handling of the postwar period in Japan began on September 5, 1945, when the General Staff issued orders that all enemy war equipment should be destroyed, except for new or unique material which was to be kept for later examination.

On October 30 the Joint Chiefs of Staff (JCS) cabled commanders in the Pacific area and in China, amplifying on these orders and instructing U.S. personnel to seize all Japanese facilities related to research in atomic energy.

When General Groves received this cable in Washington, he reviewed its contents with an officer in his charge. Groves knew of Japan's five cyclotrons and wanted to make certain they were "properly secured, in accordance with the JCS cable, but not destroyed. Yet, in the light of what followed, it's evident that I did not make my intention clear."

The errors then began. On November 7 a message went out from Groves's office in the name of the secretary of war (but without his seeing it) to General Douglas MacArthur in Japan. The cable "ordered the destruction of the five cyclotrons in which we were interested, after all available technical and experimental data had been obtained from them."

The next anybody knew, on November 24, the office of the Joint Chiefs of Staff received a cable from MacArthur's headquarters in Tokyo informing the Joint Chiefs that the cyclotrons had been seized on November 20 and that their destruction had begun on the twenty-fourth. Copies of that cable were sent to nine different people, one of whom was General Groves.

"The destruction of the cyclotrons should have come to an abrupt halt at this precise moment," writes Groves, but it did not.

Although MacArthur's cable referred only to the original JCS cable which had ordered seizure but no destruction of atomic research facilities, no one in MacArthur's office seemed to make the differentiation. Compounding the error was that of the nine people who had been sent copies of MacArthur's cable, "not one—myself included—actually saw it," Groves states.

All nine copies had been "noted, initialed and filed by subordinate staff officers, in spite of all the detailed instructions that existed in every headquarters to prevent just such an occurrence [not showing a message to a person in authority]."

Groves ascribed the problem in his own office to confusion that "stemmed at least in part from the fact that the officer handling this matter was fairly new to the project, and was unaccustomed to our way of operating."

The destruction of the cyclotrons simply went forward without anyone in the nine offices or anybody on MacArthur's staff questioning the action being taken. The first anybody knew of a problem was when, on November 28, another cable from the General Staff arrived in MacArthur's headquarters requesting that one of the cyclotrons be shipped to the United States for examination.

The bewildering message was brought to MacArthur's attention. MacArthur in turn immediately cabled General Dwight David Eisenhower to inform him of the discrepancy in orders his office had been receiving. Before Eisenhower responded on November 30, the cyclotron story hit the newspapers. The furor erupted immediately.

Now there began a back-and-forth argument about who was at fault and how had it happened that such important equipment had been obliterated. MacArthur maintained that his cable of November 24 alerting nine different offices to what was about to happen showed he had tried to bring the problem to the attention of others, but he had heard no further word. The Joint Chiefs of Staff concurred that MacArthur "had acted correctly and indicated that the situation resulted entirely from a failure to co-ordinate outgoing messages in Washington."

What then transpired is an incisive lesson in handling the aftermath of error. To deal with the uproar and what was becoming "a very bad public relations situation," the War Department proposed that in response to the mounting criticism Secretary of War Robert P. Patterson should issue a statement pointing out that the action had been taken as part of the policy to prevent Japan from engaging in any war-related activity. The recommended announcement read in part: "In order to ensure peace for generations to come we desire to eliminate to the maximum extent possible, the Japanese war-making potential. While it is recognized that a cyclotron may be used for scientific research in other fields, it is essential to the carrying out of atomic bomb research which our government believes should be prohibited to naturally belligerent and dishonest nations."

In other words, the War Department was suggesting that the secretary of war lie about the blunders that had led to the destruction of the valuable cyclotrons.

General Groves did not agree and prevailed on Secretary Patterson to be honest about the mistakes. The secretary finally issued this explanation:

> General MacArthur was directed to destroy the Japanese cyclotrons in a radio message sent to him in my name. That message was dispatched without my having seen it and without its having been given the thorough consideration which the subject matter deserved. Among other things, the opinion of our scientific advisers should have been obtained before a decision was arrived at.
>
> While the officer who originated it felt that the action was in accord with our established policy of destroying Japan's war potential, the dispatch of such a message without first investigating the matter fully was a mistake. I regret this hasty action on the part of the War Department.

When this statement was issued to the press, a remarkable thing happened. "The press as a whole seemed quite surprised by this frank and open admission of error," Groves states. "As a result, this incident quickly lost its news value and the clamor soon subsided."

For General Groves, there were many lessons in the entire episode. One was that a leader, in or out of military service, "must always make his intentions unmistakably clear to his subordinates; I did not do so in this case." Also, "even the most successful and competent organization will not continue automatically to produce perfect work when large numbers of new people are brought into it without previous experience or training."

But finally, and most important for the general, there was a lesson to be learned here about dealing with error. Concluded Groves: "Over all these other lessons looms the basic truth that was demonstrated here again, that honest errors, openly admitted, are sooner forgiven."

DO AS I SAY, NOT AS I DO

He was the new Republican clerk in the Illinois House who sent a memo reminding his staff about the need for accuracy in their writings.

His memo had a problem, however. Actually it had nine problems. When his directive calling for accuracy was distributed, it was found to have grammatical, typographical, and spelling errors—nine in all.

When the Democrats in the Illinois legislature happily made the malformed memo public, the clerk, Tony Leone, moaned in defense, "Can a guy have a classic snafu, or can a guy have a classic snafu?"

THE "BEST-LAID PLANS" DEPARTMENT:

THE LESS-THAN-PERFECT PICTURE POSTCARD

The Nazis had a reputation of being coldly efficient, but a picture postcard published two years after Hitler came to power disproves that estimation. In 1935 Nazi government officials selected a blond, rosy-cheeked boy to pose for a photograph with Chancellor Adolf Hitler as an example of a purely Aryan child. The picture subsequently was made into a postcard, and hundreds of thousands were sold throughout Germany—all before the child's true racial origin was learned. The Nazis discovered, too late, that the boy was the grandson of a Rabbi Wedell, of Düsseldorf.

CHAPTER V

Foul-up: Error in Business

A former treasurer of Harvard once told me that he had two golden rules for managing the Harvard portfolio. One: Never consult the economics department. Two: Never consult the business school.

—Paul A. Samuelson,
Nobel Prizewinner in economics

The business world prides itself on its efficient management. No waste here. No boondoggles. No lost millions on poorly conceived or executed projects. At least that's the way most business people would characterize industry and commerce in comparison to, say, government. After all, the business person is engaged in making a profit, and none of the laxity, indifference, or inefficiency that supposedly occurs in government would be allowed in the commercial setting.

How, then, do business and industry explain the Edsel? Or the stock market crash of '29? Or the fact that there are hundreds of thousands of bankruptcies each year?

For instance, consider the case of the Bechtel Group, the large and powerful company once headed by Secretary of State George Shultz. Involved in finance, construction, and engineering, Bechtel was chosen

137

to build the Washington, D.C., subway system. Originally estimated to cost $793 million for 100 miles of track, the system, when Bechtel finished, cost $6.6 billion. And then there was the time at the San Onofre, California, nuclear facility that Bechtel installed a 420-ton reactor vessel backward. And the time in 1974 when it was sued for $300 million—and eventually paid out $14 million—over the breakdown of the Palisades nuclear generator because Bechtel did not "prevent errors in design and manufacture of equipment and components." And just to show that it was not committing these costly errors only at the expense of Americans, Bechtel built the city of Jubail for Saudi Arabia at a cost overrun of a mere $36 billion. (The *New York Times* estimated that what was to have cost $9 billion ran as high as $45 billion.)

Then there was the case of the nation's fourth largest bank, Manufacturers Hanover Trust Company, which in October 1979 issued erroneous data to the Federal Reserve Bank that later had to be revised. The miscalculation caused stock market losses for many people. When it became public that the bank attributed the error to mistakes in its filling out a new form being used by the Federal Reserve, a lawyer was quoted as saying, "I didn't mind being wiped out, until I found out how."

Even with all their planning and testing, their directors of finance and highly trained accountants, their payrolls of marketing experts, lawyers, engineers, and scientists, the titans of industry have exhibited much the same propensity to commit costly errors as people in other fields of human endeavor—a fact with an added dimension for business. If in commerce the profit motive is the goal, then cost overruns and faulty projections, lapses and losses, bailouts and bankruptcies are the symptoms of a business world falling far short of its ideal. This realization, however, could have an important benefit. Understanding the many possibilities for error in commerce and industry is something by which the astute businessperson could certainly profit.

WHEN THE WALL CAME TUMBLING DOWN: THE CRASH OF '29

The business world's biggest foul-up was probably the Wall Street crash of 1929 (the famous headline in *Variety* was WALL STREET LAYS AN EGG). What followed was the Great Depression, an economic disaster still enshrouding the United States a decade later, when World War II began.

How did so many act so poorly, thereby causing so many to become so poor?

One reason the crash happened was that almost everyone erred in assessing the true nature of the times and how overspeculation had sown the seeds of disaster. In fact, newspapers and books of the day are replete with giants of the corporate world, professors of economics, and leaders of government predicting that no end of the economic boom could be foreseen.

For example, Charles E. Mitchell, chairman of the National City Bank, on September 20, 1929, a month before the crash, announced, "The industrial condition of the United States is absolutely sound." On December 4, as proof, his bank offered its employees shares of the company for $200 on a four-year installment plan (at the time the shares were selling for $221 bid, $225 asked). In less than two years — by October 2, 1931—the shares had dropped to $50 bid, $50.50 asked.

Charles M. Schwab, chairman of the board of the Bethlehem Steel Company, declared in an address to the American Iron and Steel Institute on October 25, 1929, just days before the crash, "In my long association with the steel industry I have never known it to enjoy a greater stability or more promising outlook than it does today." Yet by October 1, 1931, Bethlehem Steel was cutting all employee wages by 10 percent.

Irving Fisher, a famed economist of the times, stated on September 5, 1929, "There may be a recession in stock prices, but not anything in the nature of a crash. . . . Today [the investor] obtains wide and well managed diversification of stock holdings by purchasing shares in good investment trusts." On the day he spoke, one such investment trust— Goldman Sachs Trading Corporation—closed at $110 a share. By 1931 it was trading at 75 cents a share.

But Fisher, a month and a half later, on October 16, 1929, gave a speech which made him for decades the butt of jokes and the king of error in economic projections. In his speech Fisher proclaimed, "Stock prices have reached what looks like a permanently high plateau. I do not feel that there will soon, if ever, be a fifty or sixty point break below present levels. . . . I expect to see the stock market a good deal higher than it is today within a few months." Nine days later came what the *New York Times* in a headline termed the nation's "Worst Stock Crash."

After the stock market had started to plummet, the forecasts of business leaders proved no better than before the crash.

Alfred P. Sloan, Jr., head of General Motors, at the beginning of 1931, a year which next to 1932 was probably the worst of the Depression: "I see no reason why 1931 should not be an extremely good year."

W. S. Farish, Humble Oil Company, in December 1930: "The present depression . . . will in my belief end in six months."

Walter S. Gilford, AT&T, on November 25, 1930: "As sure as I am standing here, this depression will soon pass and we are about to enter a period of prosperity the like of which no country has ever seen before."

Government leaders, however, were no better than business leaders in assessing the depth and nature of the economy's plunge.

Secretary of Commerce Robert P. Lamont, on March 22, 1931, declared that "there undoubtedly will be an appreciable decrease in the number of unemployed by mid-summer." However on June 13—mid-summer—the *New York Times* reported that LAMONT PLANS CUTS IN HIS DEPARTMENT and that the secretary was urging division heads "to reduce personnel in economy campaign."

Dr. Julius Klein, assistant secretary of commerce, shortly before his boss cut back on staff, announced that there was "a fairly good chance that the United States will be out of the current economic depression by the end of October." On June 9, 1931, in a speech to the Radio Manufacturers Association in Chicago, Klein announced, "The depression has ended. . . . In July, up we go."

Even the respected *Harvard Economic Society Bulletin* could not get its predictions right. "A depression like 1920–21 is clearly out of the question," it stated on November 30, 1929. Six months later, as the economy was picking up speed on its downward spiral: "General prices are now at bottom and will shortly improve."

It really was Calvin Coolidge who put the times in correct perspective. The former president stated on January 20, 1931, "The country is not in good condition."

But before you think Silent Cal knew what the great titans of industry could not fathom, let us remember that it was the Coolidge presidency that laid the groundwork for the overspeculation that led to the stock market crash.

On December 4, 1928, in his final message to Congress less than a year before the crash, Coolidge showed how the blanket of error can settle over government and business together. Said Coolidge: "No Congress of the United States ever assembled, on surveying the state of the Union, has met with a more pleasing prospect than that which appears at the present time." Citing "the highest record of years of prosperity," he declared, "The country can regard the present with satisfaction and anticipate the future with optimism."

The stock market crash of 1929 was so devastating that $3 billion in stock values vanished in just the last hour of trading on October 22. Thousands of traders suddenly dumped holdings at any price, with shares down anywhere from $10 to $96 in what was termed an "avalanche" of selling. Yet even then one newspaper headlined the next day —BANKERS OPTIMISTIC.

HOW GOOD ARE STOCK MARKET EXPERTS?

The stock market is the essence of American business. Here are bought and sold the shares in America's major companies. Here are the leading business thinkers and shakers engaged in making profits and building fortunes. How good, then, are the stock market experts, those forecasters who make their livings predicting the market for themselves and countless others?

For the past twenty years a financial news service, *Investors Intelligence,* has been compiling an index of the sentiment of leading stock market forecasters to determine how the "experts" see the direction of the market at the time. Here is a survey of those predictions—and their accuracy—since 1963.

On August 2, 1963, the Dow Jones industrial average was 689. Of those surveyed, 78 percent forecast a drop in the market, and another 14 percent were gloomy enough to see a bear market ahead. Only a small 8 percent were bullish about the coming months. What really happened over the next twenty-one months? The market spurted upward 250 points.

A decade later, on January 12, 1973, the Dow Jones average stood at 1,047, a near-record level at the time. Of the experts surveyed, 62 percent were optimistic about the market's future, 15 percent pessimistic. Result: Stocks fell 470 points in twenty-three months.

Another decade later, in the first months of 1983, the stock market roared to its highest level in history, going over the 1,200 level. The rise had started in August 1982. What had the experts been predicting that previous summer? In June 1982, with the Dow around 800, approximately 55 percent of the experts were bearish, and another 25 percent expected a downward correction. Only 20 percent were optimistic.

Among the many pessimistic advisers was Joseph Granville, a leading investment counselor whose bearish pronouncement in January 1981 had sent the stock market into a one-day thirty-point slide. Wrote Granville in 1982 on the eve of the largest stock market advance in history: "Something is terribly wrong. Sell everything." By the end of the year, of more than fifty investment letters tracked by *Hulbert Financial Digest,* the Granville letter was last for 1982—its recommendations down by 25.6 percent.

THE LAST STRAW POLL

Polling—as used in political contests and to determine consumer interests and needs—has received increased attention in recent decades as more effort and money have gone into making them as accurate as humanly possible. But *humanly possible* often means just "possibility for human error."

The most glaring polling error is one that has not been topped for close to fifty years now. This is the polling mistake made in 1936 by the *Literary Digest,* which sought to determine the winner in the presidential race between Franklin D. Roosevelt and Alfred M. Landon. The *Digest,* a leading magazine of the day, had made a concerted effort over an extended period of time to poll nearly 2.5 million voters. Four days before the election the magazine announced the results of its poll: The victor would be the Kansas Republican Landon with a startling 370 electoral votes. The election, however, saw FDR swamp Landon with what was then the largest landslide in American presidential campaigns. FDR won 523 electoral votes; Landon, just 8.

What had happened to the *Literary Digest*'s poll? In retrospect, it was found that the magazine had relied heavily on the telephone to poll voters, but many Americans then did not have their own phones. The magazine's poll was therefore incorrectly weighted to a more affluent and thereby a more Republican sampling.

Within a year the *Literary Digest,* its reputation tarnished, went bankrupt and ceased publishing.

THE EDSEL: DETROIT'S BIGGEST FOLLY

In 1957 a new car rolled off Detroit assembly lines—and into business history as the single biggest mistake made by one of the giant manufacturers in America's most important industry.

The company was Ford, and the car was the Edsel. Named after the only son of Henry Ford I, the Edsel was the product of many years of research and planning and millions of dollars in start-up costs. Even the name was not given lightly; among those consulted for suggestions was the poetess Marianne Moore (two of her names for the new automobile were Utopian Turtletop and Mongoose Civique).

Indeed, the Ford Motor Company took great pains in developing the Edsel, the manufacturer's first complete new line of cars in a decade. Ford intended it to compete with General Motors in a section of the market where GM was strongest: the mid-size family car.

Among the examples of the care and attention Ford put into the Edsel:

—The company employed new research techniques in the consumer motivation field. Extensive psychological testing was done to determine the public's image of a family car, and Ford responded by designing the car to be the embodiment of a dream supposedly already existing in the minds of prospective buyers.

—To provide variety, Ford offered eighteen models in four series of Edsels. The autos were filled with fancy gadgets—such as push buttons for gears—and given a new styling, which included vertical grilles instead of the usual horizontal grilles. Ford also made the Edsel wider than most cars on the road and gave it an engine with higher-than-average horsepower.

—To sell and service the Edsel, Ford recruited 1,160 new dealers, many of whom were formerly with GM and Chrysler, and made them handle only the one line of Edsel cars—an innovation in the structure of auto dealerships. Many of the Edsel dealerships were the best in a community.

—To promote the Edsel, Ford mounted an aggressive $10 million advertising campaign even before unveiling the new car, with "leaks" of information about the Edsel designed to spark interest and anticipation among potential buyers.

As a result, on the September day in 1957 when it was officially unveiled, the Edsel was big news on the evening television broadcasts and in both general-interest and car magazines. During those first days and throughout the first month showrooms were crowded with people eager to see the much-ballyhooed Edsel.

But a major problem soon became apparent. Although many came to see, few came to buy. September sales were weak, and October sales proved even weaker. November sales dropped below October's. Dealer inventories began to climb, soon shooting to a ninety-day level. Many dealers, some after investing more than $100,000 in new dealerships, found themselves bankrupt in three months.

One key Edsel dealer, who had given up an Oldsmobile franchise for an Edsel dealership in New York City, remarked at the end of November, "The Ford Motor Company has laid an egg."

Ford, seeing the pitiful sales, moved quickly to help sell Edsels. They offered sales bonuses to dealers, organized a system for dealer exchanges of models, sizes, and colors, upped the national ad campaign to $20 million, and launched huge demonstration projects in which 500,000 "prospects" drove cars. Ford even offered Edsels at substantial discounts to state highway officials just to get the cars on the road.

But little seemed to help. During its first three years of existence only 110,000 Edsels were sold (in contrast, today a model such as the small-size Escort sells more than 300,000 in one year). By 1960 Ford had called it quits and stopped producing the Edsel. The loss to the second largest auto manufacturer in the world: a quarter of a billion dollars.

What went wrong? How did Ford make such a costly, humiliating error?

First, Ford had probably overpromoted the car before the public saw it. The buildup of anticipation was not fulfilled by the product, which, instead of being the car of the future touted to be years ahead of its competitors, proved to be only superficially different inside and out (the vertical instead of horizontal grille in the front, for instance, was the butt of jokes because of the transparent attempt to be a cosmetic rather than functional advance in auto design).

Secondly, the Edsel had possibly been in the planning stages too long. The planning for the car had begun in the early 1950s during an economic boom, but when the auto was finally introduced in 1957, the economy had begun a downturn, and the expanding consumer credit and rising incomes of the early fifties had been halted. Instead of heeding the change and introducing an economy car in response, Ford went ahead with a higher-priced Edsel.

Thirdly, the market research had been based on the wrong assumptions. The questioning of consumers had centered on cars as status symbols and as extensions of owners' egos, but no questions had addressed such issues as maintenance costs, interest in gadgets, and functional aspects of design.

Fourthly, the timing of the introduction of the Edsel was a mistake. The Edsel appeared just a month earlier than the 1958 models and competed with dealers trying to unload discounted 1957 models. In comparison, then, the new Edsel looked even more expensive than it was, and its price generated considerable consumer resistance.

Fifthly, because the Edsel came in so many models, sizes, and colors, dealers often did not have the right combination the buyer wanted, and because dealers were new, with no credit at banks based on past sales, they found they could not freely order more cars before selling the Edsels on their lots.

Finally, of the Edsels that were bought, it has been estimated that half of them were defective in one or many aggravating ways. Doors wouldn't close properly. Horns stuck. Hubcaps dropped off. Paint peeled. Transmissions froze. Brakes failed. And the famed push buttons that had replaced hand gears wouldn't respond to pushing.

In reviewing the sad story of the Edsel, *Time* magazine commented, "It was a classic case of the wrong car for the wrong market at the wrong time."

Ironically, there were two silver linings that came out of the Edsel fiasco, showing that error can produce some good. Although the Edsel was a failure for Ford, the company was able to take advantage of Edsel production facilities to produce a much more successful car—the Falcon—a year ahead of schedule. Also, owners of Edsels have seen the value of their cars escalate over the years among classic car buyers, and the buying and selling of well-preserved Edsels is a lucrative market. At least those Edsel owners who held on to their lemons have been able to reap tidy profits.

But the Ford Motor Company's colossal mistake with the Edsel has left the Edsel name as synonymous with error and failure. As one wag noted at the time Edsels were being manufactured but not bought, only one case of an Edsel's being stolen was on record.

OTHER DETROIT FOLLIES

The Edsel was not the only time American car manufacturers committed a costly faux pas. In the 1970s and early 1980s Detroit let Japanese car manufacturers make serious inroads into their business. In September 1980 *Time* magazine, in a cover story on "Detroit's Uphill Battle," commented that "the auto industry's problems rest with Detroit's managers, who failed to plan for a new-car market after the 1974 oil embargo." *Time* cited GM executives who "publicly declared that Americans would always demand full-size cars." The magazine pointed out that in order to boost quarterly sales figures, Chrysler during the 1970s "pushed questionable products onto the market on a near panic basis. One result: the 1974 Dodge Aspen and Plymouth Volare have the dubious distinction of being two of the most recalled cars in history." And the Ford Motor Company was shown once again to be behind the times. Despite pleas from then Ford president Lee A. Iacocca (he later became head of Chrysler), Ford chairman Henry Ford II "refused to give up the big profits in building big cars." The result? "As a consequence," wrote *Time* in 1980, "Ford today has the fewest small economical models" and its big models were "selling poorly," making Ford "an even sicker company" than the nearly bankrupt Chrysler.

The Edsel lives on in Detroit.

DETROIT V. JAPAN

How much difference is there between American and Japanese customers toward flaws in their cars? The *Time* magazine Tokyo bureau chief, in an accompanying article to the above-mentioned *Time* story on Detroit's car woes, noted that the Japanese are meticulous not only about the cars they produce but about the cars they buy. "Customers will refuse delivery of a new car that has a stray smudge of grease or a crooked seat seam." When Chevrolet tried to deliver its first Citation X car to Japan, the Japanese gave the GM company a list of 105 defects that had to be corrected before the car could be sold in Japan.

A YEAR TO RECALL

One year—1977, to be exact—so many defects were discovered in both U.S. and foreign cars, that automakers had to recall more cars than they produced. While Detroit & Company manufactured 9.3 million vehicles that year, because of mistakes in design or workmanship, automakers were forced to recall a total of 10.4 million cars.

"FLY ME, I'M BANKRUPT":
WHEN THE EDSEL FIASCO WAS TOPPED

While the Edsel has come to be synonymous with gigantic corporate faux pas, another debacle in American business actually dwarfs the Edsel. During the late 1950s and early 1960s General Dynamics lost $425 million on its project to build the Convair, a commercial jet airliner. This sum was twice the amount lost by the Ford Motor Company on the Edsel. Indeed, the huge General Dynamics (a $1 billion corporation with 100,000 employees at the time) suffered the loss of one-quarter of its entire net worth. Its mistake with the Convair has been termed "the largest loss that had then ever been sustained on one single commercial manufacturing project."

In a series of articles in *Fortune* magazine in 1962,* author Richard Austin Smith blamed the fiasco on the company's top management, which grossly underestimated costs and incorrectly assessed the market for the airplane. It also did not help that General Dynamics officials had to deal with Howard Hughes or that Hughes had initially suggested the project.

The first problem was that although Convair, the largest of nine divisions in General Dynamics, had experience as a builder of military airplanes and missiles, it was not a leading manufacturer of domestic airliners. In those days the leaders were Douglas and Boeing, which were then moving aggressively into the construction of jet airliners. But when Howard Hughes, who owned TWA, suggested to General Dynamics that it, too, build a jet airliner and that he would buy it for his company, the officials at General Dynamics grabbed at the idea as a way to enter the growing market.

The heads of the Convair division erred from the beginning, however. Their original estimates were decidedly optimistic: They projected the sale of 250 jet airliners (the Convair 880) over ten years, $1 billion in sales, and at least $250 million in profit. Even if everything went wrong, they felt, they stood to lose at the most $50 million. Eventually everything did go wrong, and the company lost far more than $50 million.

There were some voices within the company questioning the reasoning and the figures. The vice-president in charge of the company's

*An account of the Corvair fiasco can also be found in *World's Worst Aircraft* (St. Martin's Press, 1976), which tells about a number of blunders in the sky.

military business cautioned that on the basis of his experience, the cost estimates given by engineers were never accurate; they always seemed to be half the final costs. Other executives doubted either that the level of sales of jet planes would be reached or that Howard Hughes was a reliable enough customer on which to base such effort and investment. Overlooked by most of the company officials was the fact that although Delta and KLM had taken options, only the unreliable Hughes and his TWA were prime customers.

It also did not help that during this time a power struggle was going on at the top of General Dynamics that eventually left a division like Convair operating on its own.

The errors became apparent almost at once. Refiguring their costs, the heads of Convair realized that the original estimate of planes needed to be sold to break even (sixty-eight) was too low. Next one of their intended customers, KLM, notified them it was dropping its option. Then Howard Hughes notified Convair he would not let it sell its plane to any airline that was a direct competitor of TWA. Also, he wanted a plane with five-abreast seating, while United Airlines, a possible customer, insisted on six abreast.

The fact that the original cost estimates were wrong then began to emerge. An employee in the purchasing department made an amazing discovery: The parts of the plane General Dynamics intended to buy rather than to build itself (the engine, parts of the airframe, the radio transmitter, etc.) would cost more than $4,250,000—which was the intended sale price of the entire plane. The employee, who recommended that the project therefore be scrapped in its entirety, was fired.

Then United Airlines, which at one point had agreed to buy the Convair 880 model, opted for the new Boeing 720, a plane that allowed for six-abreast seating and offered the size and range United wanted.

Boeing's development of the 720 doomed the Convair 880. Since the Convair 880 did not have the long range of Boeing's first jet airliner, the 707, or of the Douglas DC-8, General Dynamics had decided to go after the market for the medium-range jet. The 720 now did not even leave the Convair with this market.

General Dynamics still had Howard Hughes as a customer—it thought. But Hughes did not have the financing when the first planes reached completion. With eighteen more in production, Convair officials decided to stop until Hughes could buy them. Even this cautious step proved wrong. Months later, when Hughes got his financing, the Convair people discovered—incredibly—that they had no record of exactly where in the production process each of the eighteen planes had been. Questions about the extent of the wiring, the status of modifications, the degree of engineering changes that still had to be made—all this and more took expert analysis and time to figure out. But the difficulty didn't stop there. Expenses were run up now to retrain people and hire new employees to replace those let go when production had been halted. And then there was the cost of repairing the water damage to some of the planes, left for months sitting in the open.

Finally, in December 1958, the first Convair 880s were ready. But so, too, were new troubles. American Airlines, a potential customer, now decided to buy the Boeing 720. Desperate, the Convair people came up with a new idea: a new model, the 990, which would have a General Electric fan engine and would be able to fly at 640 miles per hour, making it the fastest commercial airplane in the world. American Airlines agreed to buy this plane, but only in a deal that involved General Dynamics' taking twenty-five old DC-7s in partial exchange and guaranteeing the speed and other advanced characteristics of the plane. Convair agreed, committing itself to sell the 990 at a cost of $4.7 million apiece.

The only problem was that Convair really did not know what the new plane would cost. There was no time to build a prototype, nor would the budget allow expensive wind tunnel testing. As a result, when the first 990 was ready for flying, it was found to be slower by 30 mph than originally conceived. American Airlines, which wanted to use the 990 to tout it as the world's fastest airliner, lost interest in the plane and was able to get out of its contract for the full purchase.

Then, in January 1961, when American Airlines first did use the 990, it was found to have numerous problems. Again, Convair had to undertake costly adjustments in the plane. The cruising speed was eventually brought to within 10 mph of the original goal, but new orders from the airlines did not follow. Over the next decade the major airlines bought and flew few 880s and 990s. The Convair became more a secondary plane used by charter companies than the proud mainstay of the major airlines that had been envisioned for it.

The Convair is not the only major mistake in the aviation industry. Douglas, one of the prime competitors of General Dynamics at the time, was eventually taken over by McDonnell because of its own mistakes and runaway costs on other programs. Lockheed, which lost $100 million and almost went bankrupt over the Tristar, had to be saved by the federal government. Boeing, the winner in the battle with the Convair, once experienced a year in which the company did not receive one order and had to pare its staff from 84,000 people to 21,000.

But the Convair project had a built-in problem from its inception. Seen in retrospect, General Dynamics had difficult, if not impossible, sales figures to reach just to break even. Thus, the Convair was doomed to failure all along.

"NOBODY'S PREFECT"

"Nobody's Prefect." That's how the Chesapeake & Potomac Telephone Company headlines a notice periodically sent out with its bills calling attention to mistakes in telephone book listings.

The notice continues: "Just as you may have missed the misspelling of 'perfect' in our headline, we let some errors get by in checking (and rechecking) your Baltimore Area telephone directories."

Issued several times a year, this message is followed by a listing that runs for several pages.

One such mistake occurred in 1981, when officials of Joseph A. Bank Clothiers, Inc., asked the telephone company to drop the word *Inc.* from the store's listing in the Yellow Pages. The chain of clothing stores had been acquired by Quaker Oats, and the *Inc.* was no longer appropriate to the name. When the clothing company officials later checked the new Yellow Pages, however, they were horrified to see that their instructions had been carried out to a *T.* There in the phone book for all to see was no longer "Joseph A. Bank Clothiers, Inc.," but—as had been requested—the entry "Drop inc."

Telephone company officials in Atlanta, where the error had been made, were embarrassed by the mistake, especially when the Associated Press picked up the story and sent it out by wire service to the nation's newspapers.

WHEN THINGS DID NOT GO BETTER FOR COKE

In March 1982 a Coca-Cola bottler in Cookeville, Tennessee, uncorked a new contest in which players had to acquire letter-embossed bottle caps spelling out "Home Run." To make certain there was not a flood of winners, the game was conceived to make *R*-lettered caps few and far between. In fact, the original odds were a mere 1 million to 1.

But as the result of a mistake made by the printer, the *R*-lettered caps began popping up more than planned. In fact, 18,000 times more than planned. Suddenly scores of people were turning up with winning entries—and collecting $100,000 in awards before the local bottler, citing contest rules allowing it to back out in the case of such an error, apologized and put a cap to the contest.

One of the many angry contestants, however, voiced his discontent. Quoted in a *Newsweek* magazine story on the mistake, he said he was going to "let everybody know that Coke is not the real thing."

TWO FAMOUS BUSINESS INVENTIONS TURNED DOWN BY CORPORATIONS

Businesspeople are supposedly always on the lookout for innovations that will help them get ahead of their competition. After all, thinking ahead, sensing a need, and filling it are how most of the fortunes in the business world are made. So it stands to reason that the corporate world should always be on the alert for and receptive to fresh ideas and new inventions. Right? Wrong. Two famous illustrations:

—Alexander Graham Bell, the inventor of the telephone, an invention without which the business world of today could not even begin to function, was hard pressed to find a major backer. In 1876, the year he patented the telephone, Bell approached Western Union, then the largest communications company in America, and offered it exclusive rights to the invention for $100,000. William Orton, Western Union's president, turned down the offer, posing one of the most shortsighted questions in business history: "What use could this company make of an electrical toy?" There is, however, some poetic justice in the corporate world. The telephone eventually consumed the lion's share of the market that Western Union had dominated.

—Chester Carlson worked diligently for years to interest various companies in his invention. Kodak turned it down. Twice the IBM Corporation studied the invention and twice rejected it, once on the advice of the Arthur D. Little Market Analysis Company. The Haloid Corporation, a nearly bankrupt company, finally decided to risk putting money into the invention to see if it could reverse company prospects. It did. Carlson is the inventor of xerography, and the Haloid Corporation is today the Xerox Corporation. IBM, although doing very nicely with computers, has never been able to equal Xerox's dominance of the office copier field.

SOMETHING TO THINK ABOUT

In the United States today:

>   *At least 50,000 businesses go bankrupt annually, most often because of owner or operator error and foul-up.
>   *And not just small companies but giant companies go out of business or have to be rescued from their mistakes. Some of the corporate giants that in recent years needed to be saved from bankruptcy: Penn Central Railroad, Lockheed, W. T. Grant & Company, Chrysler.
>   *Even entire industries, such as steel, electronics, automobiles, and optics, have failed to plan for and combat grave problems that threaten their future.

JUST DON'T MENTION THE WORD *CORFAM* AROUND DU PONT

The story of the Du Pont corporation and Corfam is a story of an on-again, off-again, "I-don't-even-want-to-hear-your-name-again" romance.

Du Pont, a giant corporation which has successfully developed and marketed a number of synthetic fibers, once set out to develop a substitute for leather. The result was Corfam, and Du Pont saw it as a savior for the shoe industry. The problem was that the consumer originally did not see it that way. Shoes made of Corfam did not sell, even with Du Pont's prodding and special marketing strategies. For instance, Du Pont first put Corfam into expensive shoes to enhance the image of the artificial material before using it in cheaper shoes.

When little seemed to make the marketplace respond to Corfam, Du Pont finally decided to sell its multimillion-dollar mistake. The sale recouped only a small part of the company's expenses.

However, soon the sale only served to underscore Du Pont's frustration and losses, for eventually demand for Corfam began to rise. What made the difference were stiff increases in the cost of leather and with it the price of leather-made shoes. As a result, Corfam sales eventually shot up. Thus, Du Pont made not one but two massive mistakes with Corfam. The company first tried to market Corfam when consumers were not interested; Du Pont then sold its innovation to another company when interest in and need for the product was about to rise.

E.T., DON'T PHONE HOME

Alexander Graham Bell's telephone has proved to be a vital tool for the world of commerce. The current proliferation of the toll-free 800 number over which to sell products is the latest example of the invention's usefulness to business.

It may therefore come as a startling surprise to many businesspeople to learn how widespread are telephone billing errors for commercial companies. In fact, Sylvia Porter, the financial columnist who reported on this in her syndicated column of November 5, 1982, called the statistics a shocker: Of more than 30,000 business telephone accounts monitored over the previous fifteen years, she wrote, "billing mistakes were uncovered in a full 45 percent of them!"

Porter went on to cite a telephone consulting firm that estimated that more than nearly half of all businesses—particularly the small business customer—are overbilled on phone services.

As a result, Porter strongly urged her business readers—as well as consumers—to check their monthly telephone bills carefully.

CLOSE ENCOUNTERS WITH A CRITIC

A prediction about a movie's box-office potential was once so far off the mark that it almost cost the movie company millions of dollars in lost stock value. This occurred in connection with *Close Encounters of the Third Kind,* one of Hollywood's great success stories.

When *Close Encounters* was being shown in sneak previews, a business reporter for *New York* magazine slipped into the Dallas theater where it was being shown in advance to a select group and wrote the first review of the much-anticipated movie. Appearing in the October 31, 1977, issue of *New York,* the article, "An Encounter with 'Close Encounters,'" by William Flanagan predicted that the film would be a huge commercial failure.

"In my humble opinion, the picture will be a colossal flop," Flanagan wrote in his article. "It lacks the dazzle, charm, wit, imagination and broad audience appeal of *Star Wars*—the film Wall Street insists it measure up to. . . ."

Accompanying the article was an illustration of a spaceship landing on earth and laying an egg.

The stock of Columbia Pictures, producers of the movie, for which the company had high hopes and high profit expectations, dropped $1.375 a share on large volume the Monday the magazine appeared. On Tuesday the influx of sell orders was so high for an orderly handling that the New York Stock Exchange refused to permit trading in the stock to open until noon. When it did, the stock dropped another $1.50.

But by now *Time* magazine had appeared with a rave review by Frank Rich, one of its film critics at the time, who also had sneaked into the Dallas screening. He called *Close Encounters* "a dazzling movie," one that "reaches the viewer at a far more profound level than *Star Wars.*"

Eventually the movie and the stock soared, but not before Flanagan's mistaken assessment of the film's commercial potential had cost millions of dollars in losses on the stock exchange.

THEY LOST BIG AT THE MOVIES

The motion picture business is one industry where mistakes can result in lost millions in a brief span of time. Here are some of the movies that have lost more than $10 million. Leading such a list, and now almost synonymous with mistakes made in the movie business, is *Heaven's Gate,* which cost more than $40 million but closed after its first week of showings during 1981. Among other major business blunders in Hollywood history and how much the films lost:

1. *Waterloo* (1969)—$23.6 million
2. *Darling Lili* (1970)—$18.7 million
3. *Fall of the Roman Empire* (1964)—$18.1 million
4. *Cleopatra* (1962)—$18 million
5. *Sorcerer* (1977)—$16.1 million
6. *Mohammed Messenger of God* (1976)—$15 million
7. *Dr. Doolittle* (1967)—$13.8 million
8. *The Greatest Story Ever Told* (1965)—$13.1 million
9. *Star!* (1968)—$10.8 million
10. *Mutiny on the Bounty* (1962)—$10.2 million

WELL, AT LEAST THEY SPELLED HIS NAME RIGHT

Automakers, both large and small, seem to have their recurring troubles. John De Lorean was no exception. The former GM executive who tried to start his own car company only to go out in a cloud of drug dealing and bankruptcy was once praised by *Newsweek* magazine, which headlined its article on him "De Lorean's Dream Is Selling Well." The date of the article was January 11, 1982. Slightly less than a month later—on February 2, 1982—the *New York Times* reported that De Lorean had placed his Belfast car plant in receivership. The end of the company soon followed.

THERE'S NO MONOPOLY ON ERROR

Monopoly, the popular board game that is still selling briskly decades after being introduced in 1935, was actually first turned down unanimously by executives at the Parker Company. Although this manufacturer of board games eventually did market Monopoly, the initial Parker reaction—and the one that almost doomed America's most popular board game—was that it contained "52 fundamental errors" which would prevent Monopoly from ever being successful.

I learned early in my business career that .400 was a great batting average, but it wasn't until much, much later that I realized this applied to the personal side of things as well as the professional. I used to expect perfection from myself and perfection from those around me. It didn't work. . . . With each new year of experience, I have accepted the fact that I will always make mistakes, that I will always commit errors of judgment. . . . This was an important lesson for me to learn, and once I accepted the fact that .400 was a batting average to be proud of rather than ashamed of, I began to hit more home runs.

—Henry C. Rogers,
of the Hollywood public relations firm Rogers & Cowan
in his autobiography, *Walking the Tightrope*
(William Morrow, 1979)

THE $300,000+ OVERSIGHT

The International Business Machines Corporation has nearly three-quarters of a million stockholders. So when an error occurs in a communication with stockholders, the sheer size of the company makes the correction of a mistake extremely costly.

Such a costly foul-up occurred in the mailing of a proxy solicitation letter in spring 1982. Because of a technical error, IBM had to void a vote of the stockholders at the annual meeting, held that year on April 26, and send out a new round of proxy statements in June to its 740,000 shareholders.

The vote taken at the April 26 meeting was in error because it was discovered too late that the proxy solicitation did not describe the stockholders' right to receive money for their shares if they opposed an amendment calling for the elimination of preemptive rights. These rights permit shareholders to retain their proportionate interest when new stock is issued.

The mistake, said an IBM spokesman, was made by IBM's in-house counsel while preparing the initial proxy.

The cost to IBM because of an error by its high-priced legal talent: $300,000 in postage alone.

THE $100,000 *R:* WHEN *TIME* ERRED

While some companies shrink from the cost of fixing an error in a product already in the marketplace, a few rare companies do not mind spending large sums of money to correct even a small error before it reaches the public.

One such company is Time Inc.

The March 21, 1983, issue of *Time* magazine featured a cover story on Lee Iacocca as "Detroit's Comeback Kid." But also part of the cover was a blurb on Henry Kissinger's "New Plan for Arms Control."

The problem was that someone at *Time* had lost control of *Control,* and the word came out without the *r* as *Contol.*

By the time an employee discovered the error more than 200,000 covers had been printed.

What to do?

Time executives did not consider any other action but one: The presses were stopped, the misspelled word was corrected, and the covers with the mistake were withdrawn.

In its sixty years *Time* had not had any similar foul-up of its cover, according to managing editor Ray Cave, but he said *Time* would never let such a mistake go uncorrected. "Of course you *could,* but we *wouldn't,*" he stated, citing as one of the magazine's guiding principles the fastidiousness of its founder, Henry Luce, about typographical errors. "It's a legacy he left us all," Cave said.

The legacy is not a cheap one. To put the *r* back in *Contol* cost *Time* $100,000 and delays of a day for 40 percent of its newsstand copies. But then *Time* magazine so rarely errs.

THE SHOEMAKER'S CHILDREN SYNDROME

In 1979 the Allied Roofing and Siding Company of Grand Rapids, Michigan, was engaged in cleaning snow from roofs in the area to prevent damage or collapse from the weight of heavy snows. But guess what roof did collapse from the weight of snow? The roof over the Allied Roofing and Siding Company.

SAY THAT AGAIN?

Since commerce crosses international borders, the business world is susceptible to the many errors that are made in translations and between cultures. One major foul-up in this regard happened to General Motors. With much ballyhoo the giant company introduced its compact car the Nova to a Latin American country only to discover too late that in Spanish the car's name is *No va,* which means "It doesn't go."

But even when companies have taken pains to translate a product or directions, errors occur. Sylvia Porter, nationally syndicated business writer, offered some examples of this in a September 1981 column:

—A company had to translate into Italian its English instructions on the use of a dentist's drill operated with a foot pedal. The translation later had to be corrected because it came out "The dentist takes off his shoe and sock and presses the drill with his toe."
—When an instruction booklet on a computer was translated for use in Indonesia, the company learned too late that the term *software* came out as *underwear, tissue,* and *computer junk.*
—A language expert hired by a corporation to handle a translation from English into Japanese wound up

creating a hitherto unknown Japanese character that meant "he who envelops himself in ten tons of rice paper."

THE ALLIGATOR SHIRT ISN'T

One of the most successful merchandising efforts in the clothing industry is the alligator line of leisure wear. These are shirts, shorts, sweaters, even socks with a small green alligator sewn onto them for a true example of conspicuous consumption. As a trademark it rivals the Coke bottle or the Camel cigarette. The only problem is that the insignia is not an alligator but a crocodile. The shirts were created by René Lacoste, a French tennis star of the 1920s who was called *le crocodile.*

Actually alligators are found only in the southeastern United States and in part of China. In Africa the animal slithering along riverbanks is the crocodile. Even the animal found in the Amazon area of South America is neither a crocodile nor an alligator, but a caiman, which is smaller than an alligator.

So please don't make the error of calling them alligator shirts. They are really crocodile—or caiman—shirts. Either way, though, the price is still the same: high.

THE GREATEST ENEMY OF QUALITY

TDK (it stands for Tokyo Denki Kagaku Kogyo or Tokyo Electrical Chemical Industries) is the world's largest producer of ferrite cores, which are used in, among other products, videotape recorders. The company's Akita plant is considered one of Japan's showcases for an automated factory. It is totally computerized with no human production workers. In its annual report, TDK president Fukujiro Sono wrote why the company had striven to remove people from the factory: "Human mistakes are the greatest enemy of quality."

THE "LOST IN TRANSLATION—AND NEVER FOUND" DEPARTMENT:

HOW A LIVELY MISUSE OF LANGUAGE MAKES FOR A DEAD LETTER

One of the major causes of error is language. The Tower of Babel crumbled under the confusion among tongues.

To show how difficult it is to put the idiom and thought processes of one culture and language into that of another, I submit the text of a letter received in 1981 by a Washington, D.C.-based international organization. The letter was from a journalist in India. Although the writer wrote in English, it is obvious his mind was elsewhere, in another language. Here is the letter, with not one word, punctuation mark, or capital letter changed:

Sir,

SUBJECT: SINGLE REVIEW COPY OF YOUR PUBLICATIONS AND BOOK REQUESTS FOR TO THIS END. INFORMATIONS REGARDING YOUR NEW RELEASES TIME TO TIME SUPPLY OF URGENTLY TO THIS END.

Single review copies of your publications and books ever been requested to this end. May I expect them single copy of titles of your interests choice selection and final approval to this end to your earliest convenience. Paper-backs your back-dated titles ever-green ones will do and suffice for the present. Review copies of your Major Books Publications all the more are welcome to this end today in the country as never before to add to your trade prospects too in the country today. Reviews will appear simultaneously in Journals Magazines Sunday Magazine Sections Columns of wide circulation varied groups of audience publicity and promotion.

Clippings of published reviews in duplicate will follow as usual. Notifications regarding your new releases time to time are so urgently requested this end to enable me to suggest time to time titles need immediate review here today in the country as never before in all interests in your business interests too to enable me to further the cause in hand here today. May I as such under the circumstances so request you and hope to be so added to your Mailing Lists so far the despatches to this end of review copies *and your catalogs complete todate are concerned* regularly.
Looking ahead   With regards   Sincerely Yours

. . . . , JOURNALIST

CHAPTER VI

Blooper:
Error in Sports

*Managers are always learning, and mostly
from our mistakes. That's why I keep a list of
my mistakes at home for reference. I used to
carry the list around in my pants pocket, but
I finally had to stop. It gave me a limp.*

—Earl Weaver, former manager
of the Baltimore Orioles,
in his autobiography
*It's What You Learn
After You Know It All That Counts*

The world of sports offers the rare opportunity to see human beings competing in a controlled environment. No matter what the sport, athletes try to reach the same goal on the same field, bounded yet freed by specific rules to which everyone must adhere, measured against statistics and records by which everyone is judged equally and without bias. Here we can see people striving to succeed, performing ever-more-prodigious feats, achieving ever-greater records. But here, too—ever-present, ever-lingering—is error.

Indeed, sports is the one field of human activity in which error can be seen most clearly. In trying to attain the highest level of physical perfection, the human being operates at his or her most vulnerable level of mishap and mistake. In fact, one sport—baseball—has from its

161

beginning kept a special statistic, clearly labeled "error," for just such mistakes during a game. Throughout the history of baseball, tens of thousands of errors have been made, and each year skilled, highly paid professionals commit thousands more. Other sports, too, are rife with examples of lapses that lead to losses.

What follows are cases in point of error at work in a variety of sports. Here are instances of both the superstar and the mediocre player bobbling and blundering. And amid the intensity and pressure of sports, the athlete is not the only one caught up in the annals of error. As shown here, so, too, are referees and umpires, managers and trainers, sportswriters and announcers, official scorers and scoreboard operators. Also owners, commissioners, and even presidents of the United States.

In many instances, all an athlete—or anyone—can do in the face of error is to grin and bear it. But an important attitude about error was expressed by St. Louis Cardinal shortstop Ozzie Smith during the 1982 National League championship series with the Atlanta Braves. In commenting after a crucial game on a teammate's baserunning blooper, Smith told reporters, "Mistakes are going to be made, but if you don't learn from it, then the mistake wasn't worth making."

Baseball—Its History

THE ERROR ENSHRINED IN THE BASEBALL HALL OF FAME

COOPERSTOWN
Here the game of baseball,
invented by Abner Doubleday,
was first played in 1839

—Highway marker, outskirts of Cooperstown, N.Y.

The Baseball Hall of Fame stands today in Cooperstown, New York, because of one of sports history's greatest errors—the commonly accepted story of the origin of baseball. Indeed, many sportswriters and announcers can still be found relating how Abner Doubleday originated baseball in 1839 on a field in Cooperstown. Yet for decades now research has shown that everything about this scenario of the origin of baseball is erroneous. In fact, to construct the Baseball Hall of Fame at Cooperstown to memorialize what Doubleday supposedly had wrought there was to erect the wrong shrine at the wrong place to honor the wrong man to commemorate the wrong date.

How, then, could so many errors occur? How could Abner Doubleday be cited as the creator of baseball and Cooperstown the site of its first game—and all seemingly enshrined by the location in Cooperstown of the sacred Baseball Hall of Fame—if it weren't true?

The story behind this great American mistake begins in the late 1800s. At that time, in response to the growing popularity of baseball, many Americans became interested in uncovering its origins. At first, and without much evidence, people simply made the claim that baseball was indigenous to America and owed none of its origin to other countries, but no one had facts with which to substantiate this. And then, in 1903, Henry Chadwick, America's first baseball reporter, who had been born in England, wrote an article in *Baseball Guide* in which he noted how much the sport seemed to derive from the English game of rounders, which he had played as a youth with "balls" and "four stones or posts in position as base stations."

Chadwick's article caused a storm. Few Americans were willing to believe such a seemingly American game as baseball could originate elsewhere.

To settle the controversy, Albert G. Spalding, a famous baseball player of the nineteenth century and founder of a sporting goods manufacturing company, proposed establishment of a committee to "learn the real facts concerning the origin and development of the game."

A seven-member committee, chaired by Abraham G. Mills, president of baseball's National League, studied the subject over several years. In 1907 the committee issued its findings: Baseball had been invented by Abner Doubleday as a youth in Cooperstown in 1839 and was clearly American in origin. The only basis for this finding, however, appeared to be the testimony of Abner Graves, who had once lived in Cooperstown and who said he remembered seeing Doubleday organize a game of baseball in a field during the summer of 1839.

Not everyone was satisfied with the committee's report. For one thing, Graves, who had been interviewed sixty-eight years after the event, was not considered the most reliable witness, and Mills, who virtually wrote the report himself, was suspect because he was a good friend of Doubleday's. But the town of Cooperstown was delighted. During following years the townspeople constructed a ballfield on the site where Abner Doubleday supposedly played his first baseball game, later erected a baseball museum beside the field, and then, in 1939 on the hundredth anniversary of an event only one person had ever said he witnessed, opened the Baseball Hall of Fame as an addition to the museum.

The unveiling of the Baseball Hall of Fame on June 12, 1939, further enhanced the myth of Doubleday, Cooperstown, and American-made baseball. Doubleday's portrait was hung above a fireplace in the building. The state of New York erected a marker at the entrance to the baseball field to cite "the birthplace of baseball" officially. The U.S. Post Office issued a stamp commemorating the centenary of baseball's founding. And President Franklin D. Roosevelt declared in a message that "we should all be grateful to Abner Doubleday. Little did he or the group that was with him at Cooperstown, N.Y., in 1839, realize the boon they were giving the nation in devising baseball."

Careful research, however, reveals a different story. Evidence now shows that baseball did originate from the English game of rounders, that the transformation from rounders to baseball was going on years before 1839, that an early, primitive form of baseball was played on fields other than Cooperstown, that Doubleday may not have even been in Cooperstown in 1839 since he was then enrolled at West Point, and that Doubleday may not have had much, if anything, to do with originating baseball (for example, after retiring from the army, he began writing his memoirs but never mentioned baseball).

One of those responsible for this definitive research was Robert W. Henderson of the New York Public Library, who presented data that, as one historian noted, "proved that all the claims made [for an American-Doubleday origin of baseball] were spurious." Among the evidence: The first printed rules for baseball appeared in 1834 in *The Book of Sports* by an American, Robin Carver, who stated that "baseball" was a change in name from "rounders"; the rules Carver cited for baseball were a direct copy of the rules of rounders published in London in 1829 in *The Boys' Own Book* and reprinted that year in America, and finally, there appeared with these rules and Carver's 1834 report a woodcut showing boys playing a game of baseball—not in Cooperstown but on Boston Common.

Even earlier references to baseball exist. A soldier at Valley Forge recorded in his diary in 1778 a game of "base." A baseball club has been found to have existed in Rochester, New York, in 1825. And there are records of the Olympic Town Ball Club's being established in Philadelphia in 1833.

Frank Menke discusses the Doubleday-Cooperstown affair in depth in *The Encyclopedia of Sports* (Doubleday, 1977). His finding: Not only did Doubleday not invent baseball, but he probably never even played it.

Baseball—Playing the Game

HALL OF FAME ERROR CHAMPS

While baseball has its Hall of Fame in Cooperstown, our Baseball Error Hall of Fame is located in Blooperstown. Here can be found the all-time error-prone players in baseball history. The men listed below hold the record for most career errors at their respective positions:

Position	Player/Years Played	Errors
First base	Cap Anson (1876–1897)	568
Second base	Fred Pfeffer (1882–1897)	828
Third base	Walter Latham (1880–1899)	780
Shortstop	Herman C. Long (1880–1903)	1,037
Catcher	Ivey Wingo (1911–1929)	234
Pitcher	James Vaughn (1908–1921)	64
Outfield	William E. Hoy (1888–1902)*	384
	Max Carey (1910–1929)	235
	Ty Cobb (1905–1928)	271

*Hoy, who was totally deaf, was called Dummy by fellow players, and he is listed as Dummy Hoy in many official baseball listings. Because of his handicap, umpires began using hand signs for balls and strikes and baserunning decisions. Hoy died in 1961 at the age of 99.

BASEBALL'S ALL-STAR TEAM OF ERROR

One can almost make an All-Star error-prone team out of otherwise All-Star players. Consider the record of error compiled by the following players, all of whom are in the Baseball Hall of Fame.

FIRST BASE—George Sisler and Cap Anson are tied with four others for leading the American League in errors at first base for the most seasons (5).

SECOND BASE—Eddie Collins, who played for Philadelphia and Chicago in the American League, holds the modern career record for most errors in a league (448). Nap Lajoie is tied for the modern record for most errors in a game (5).

THIRD BASE—Pie Traynor of the Pittsburgh Pirates, considered the greatest of third basemen, made more errors than any other third baseman in modern National League history (324).

SHORTSTOP—Honus Wagner of the Pirates holds the National League record for most errors by a shortstop (676). Luke Appling led the American League in errors the most seasons (5).

OUTFIELD—Ty Cobb, the great Detroit Tiger hitter and baserunner, was also the great error maker. He holds the modern career record for most errors made by an outfielder (271).

PITCHER—Warren Spahn stymied batters for years as a hard-throwing pitcher for the Boston/Milwaukee Braves, but he was less than successful as a fielding pitcher. In fact, he tied the record for leading the major leagues in errors at his position for the most seasons (5).

Some of these record-making performances of error are undoubtedly a function of success: The longer a player is in the major leagues, the more his exposure to and possibility for error. But these statistics also show that the great are not immune to mistake, and that we are measured not so much by our miscues as by our contributions. It's a heartening thought.

THE ONLY PLAYER TO COMMIT
MORE THAN 1,000 ERRORS

The man who committed more errors than any other player in professional baseball history—in fact, he is the only baseball player to commit more than 1,000 errors—was a shortstop named Herman Long.*

Long, who was born in Chicago in 1866, played for fifteen seasons,

---

*A Herman Long baseball card in mint condition is today worth $100. Interestingly, this is a higher figure than for most other baseball cards of his era. Error pays, sometimes.

beginning in 1889 with Kansas City, then with Boston, New York, Detroit, and Philadelphia. In all he played a total of 1,877 games: 1,792 at shortstop, 65 at second base, 19 in the outfield, and 1 at third base. During these 1,877 games, he committed 1,037 errors, a rate of almost one every game and a half or once every fifteen innings.

The only thing that could keep such a notoriously bad fielder in professional baseball was an ability at the plate, and Long, who was nicknamed Germany, was a very good hitter indeed. In fact, in 1900, playing with the Boston Braves in the National League, he led the league in home runs, hitting 12 during the dead ball era. He also had the highest frequency of home runs per at bats for the season (an average of 1 home run out of every 40 times at bat; in contrast, however, during the live ball era, Babe Ruth hit a homer once out of every 8½ at bats).

In addition to his home run hitting, Herman Long had a very respectable lifetime batting average of .279. And at five feet eight and a half inches and 160 pounds, he was fast enough to steal 534 bases.

But he really made a hit when it came to errors. Long's great feat of error was so prodigious that it makes any other career fielding performance pale into insignificance. The closest player in total errors was William Dahlen, who played more than 2,100 games at shortstop in the National League (1891–1911) and committed 972 errors. But he needed 300 more games to make 65 fewer errors than Long.

Herman concluded his career in 1904 with one game for the Philadelphia Phillies. He died five years later in Denver, Colorado, at the age of forty-three. But he left a career record of botchery that may last as long as baseball is played—and misplayed.*

PIANO LEGS HICKMAN—A MAN FOR ALL SEASONS

Another baseball player who must rank as one of the great standouts in the annals of error is Charles Taylor Hickman, better known during his playing days as Piano Legs Hickman. During the 1900 season he made 91 errors in 120 games as the New York Giants' third baseman, flubbing 2 out of every 10 chances for an .836 fielding percentage—the worst one-season performance in errors ever recorded. John Goch-

---

*Another record for miscues that may endure was achieved by the shortstop-second base combination that played for Chicago in the National League in 1885. Thomas Burns, shortstop, teamed up with Fred Pfeffer, second baseman, to commit 182 errors in 112 games. In 1886 Burns was shifted to third.

nauer, shortstop for Cleveland, committed 95 errors in 1903, but it took him 128 games to flub this many. His fielding average (.869) was therefore not as brutal as Hickman's, whose .836 fielding percentage ranks as baseball's *lowest* for a season.*

Hickman, born in 1876 in Dunkirk, New York, stood five feet eleven and a half inches tall and weighed 215 pounds, so it's easy to see why he got his nickname.

But Piano Legs made up for his fielding butchery with his bat. As a twenty-four-year-old in his fourth year as a pro during that 1900 season, he hit safely in 27 straight games (one of the longest streaks at the time) and averaged .313, with nine home runs, seventeen triples, and nineteen doubles, a powerful output for the dead ball era. His lifetime batting average, garnered over twelve years with teams in Boston, Cleveland, Detroit, and Chicago as well as in New York, was a solid .300.

In addition to playing third base, he was used sporadically as an outfielder, shortstop, even pitcher (in 1899 he won 7 games without a loss—one of the few pitchers to win more than 5 games in a season and not lose any).

Hickman died in 1934 in Morgantown, West Virginia, at the age of fifty-eight, but he lives on in a special niche in the Error Hall of Fame.

THE PLAYER WHO MADE FOUR ERRORS
ON ONE BALL

Mike Grady, third baseman for the New York Giants, holds a record for error hardly surpassed since the day in 1899 when he committed four errors on one ball.

Grady's problem began innocently enough when an easy ground ball was hit his way. Here's what then happened:

ERROR ONE: Grady bobbled the easy ground ball, enabling the batter to reach first base safely.

ERROR TWO: Grady, trying to catch the runner anyway, fired to first, but the ball sailed high above the first baseman's glove, permitting the runner to reach second base.

ERROR THREE: When the runner reached second and saw the ball

---

*In 1972 Hal MacRae, outfielder for Cincinnati, had an .833 average, but this involved the commission of only 6 errors since he did not play regularly.

was still being chased down by the fielders, he began running to third, where Grady stood. The first baseman, by now reaching the ball and seeing the runner trying to go to third, threw to Grady to tag out the runner. But Grady dropped the ball, allowing the runner to reach third.

ERROR FOUR: When he saw Grady drop the ball, the runner just kept running and headed for home plate. Grady recovered the ball and threw to the catcher, but again a Grady throw sailed over an outstretched glove, this time into the stands, and the runner scored.

The official scoring of the episode: 0 hits, 1 run, 4 errors.

THE BLUNDER THAT COST A PENNANT

The name Fred Merkle has entered the baseball history books as the player who lost a pennant for his team—not for something he did but for something he didn't do.

It was September 23, 1908, and the New York Giants and the Chicago Cubs were locked in a close pennant race. Just the day before, New York had lost a doubleheader to Chicago, thereby escalating a sudden decline that had driven the Giants from a comfortable lead into a tie with the Cubs.

As the teams took the field at the Polo Grounds that September 23, New York fans anxiously waited to see if their team could pull itself out of its tailspin and go on to the pennant many had felt confident about only a week before.

By the bottom of the ninth, with the Giants coming to bat, the score stood at 1–1. The first two Giant players failed to get on. Then, with two out, Moose McCormick reached first. Up came Fred Merkle.

Fred was a nineteen-year-old in his first full season in the major leagues. He would go on to play another fourteen seasons in the majors and live to be sixty-seven. But what was about to happen to him would dog him throughout his life.

Merkle singled, sending McCormick to third and bringing up Al Birdwell, who then smacked a line drive safely into center field. Normally this would have scored McCormick from third with the winning run; that is what happened—almost. McCormick did race across home plate, but Merkle, who was on first and who had begun running toward second as part of the play, stopped when enthusiastic New York fans began pouring onto the playing field to celebrate the scoring of the winning run. Instead of touching second to complete the play, however,

Merkle decided the better part of valor and baserunning was to head immediately for the dugout and avoid the fans.

Frank Chance, the Chicago manager and the first baseman in Chicago's famed Tinkers to Evers to Chance infield, realized Merkle's mistake and began to protest that until Merkle touched second, the run should not count. The Giants also realized the blunder. Giant pitcher Joe McGinnity quickly seized the game ball and threw it into the stands.

What then happened is a matter of controversy. Some say that a Chicago player simply got another baseball from the dugout and the Cubs used it to touch second base and force out Merkle, who at this point was not even on the field. Johnny Evers, the Chicago second baseman, later claimed that Joe Tinker was able to get the original game ball and get it to him for the out.

In either case the umpire, Hank O'Day, agreed with Frank Chance's protest. He called Merkle out for not touching second, said that the run did not count, and announced that, therefore, the game was still a tie.

The Giants now protested. They said they had won, 2–1. The Cubs in turn proclaimed they had won by the forfeit score of 9–0 since the Giants had left the field without completing the game.

Only a Solomonic decision could handle this, and National League president Harry Pulliam tried his best. He ruled the game a tie and ordered it replayed from the beginning if at the end of the season the game was necessary for determining the pennant winner.

As Merkle's luck would have it, that is precisely what happened. The Giants and the Cubs wound up the regular season with identical win-loss records of 98–55. The "tie game" caused by Merkle's mistake of not touching second base had to be replayed.

There was no joy in Mudville or in Merkle's heart the day the Giants and the Cubs met for the replay. This time the Chicago Cubs beat the New York Giants, 4–2, to win the pennant and put Fred Merkle into the Hall of Fame of Error.

The irony is that Merkle was not the only baseball player to commit such a faux pas on the base paths. In fact, just nineteen days before Merkle's lapse Umpire O'Day had witnessed another player do the same thing in a game at Pittsburgh—and at that time O'Day had allowed the winning run to stand even though the runner had not touched second base. But Merkle's mistake came during a critical game on which a pennant—and a reputation—hung.

MISADVENTURES OF THE MIGHTY

We have seen how even Hall of Famers can commit errors in record numbers. But some of the giants of baseball history have other glaring lapses in their records. For instance:

Babe Ruth: An awesome batter, with 714 home runs, he held the record for most career homers for thirty-nine years . . . but at one time he also struck out more times than any other player in baseball history (1,330 strikeouts). And on September 11, 1931, in a game with the Chicago White Sox, he grounded into a triple play.

Ty Cobb: A fierce competitor, batting champ, and once the holder of most base steals in a season and in a career, until 1982 Cobb also held the record for being thrown out the most times attempting to steal in a season (38 times in 1915).

Cy Young: With 511 victories to his credit, Young holds the record for most career wins by a pitcher. However, he also had 313 losses, the record for most losses by a pitcher (he lost 20 or more games in a season three times and once had a 13-21 season).

Hank Aaron: Slugger of most home runs in a career with 755, Aaron also holds the career record for hitting into the most double plays.

Walter Johnson: One of the greatest pitchers of all time (until recently the holder of most strikeouts over a career —3,508), Johnson holds the record for hitting batters (204) and is tied for most wild pitches in an inning in the American League (3) and most wild pitches in a career (156).

Jimmy Foxx: The great right-handed batter (he once hit 58 home runs in a season) holds the record for leading a league in strikeouts for the most consecutive seasons (7).

Roberto Clemente: The Pittsburgh Pirate batting star once struck out 4 times in an All-Star game—a record.

Sandy Koufax: The pitching ace for the Brooklyn Dodgers, with four no-hitters to his credit, had a difficult time as a batter. He holds the record for striking out the most consecutive times at bat: 12.

Joe DiMaggio: The Yankee Clipper once hit into 7 double plays during a World Series—still a record.

Willie Mays: The "Say Hey!" Kid almost said "Oh, no!" when in one World Series game he hit into 3 double plays, a record that still stands.

Reggie Jackson: On May 13, 1983, in a game against the Minnesota Twins, Jackson, playing with the California Angels, became the first major leaguer to strike out 2,000 times. Asked what this kind of record meant to him, the slugging outfielder said, "It means I did nothing but miss the ball for four full seasons."

WHEN ONE PITCH RUINED A PITCHER'S GREATEST SEASON

Jack Chesbro (1874–1931), inducted into the Hall of Fame in 1946, had in 1904 what has often been termed the finest season of any pitcher in the twentieth century, yet his error on one pitched ball in the last inning of the last game of the season cost his team the pennant and left him with a keen disappointment for the rest of his life.

In 1904, pitching for the New York Highlanders of the American League, Chesbro had a record 51 starts, hurled 48 complete games, led the league with 454 innings, struck out 240 batters, had 14 straight wins, completed 30 straight games, and compiled an astonishing 41-12 record. Chesbro's 41 wins are still the most games ever won by a pitcher in one season.

But the season came to a crashing conclusion for him on October 10, 1904, in a critical pennant-deciding game against the Boston Red Sox. In the ninth inning, with the score tied 2–2, two out, a runner on third, and an 0-2 count on the batter, Chesbro threw a wild pitch that sailed over the catcher's head. The runner scored, and Boston won, 3–2, capturing the pennant.

That pitch, which cost his team the league championship, lingered with Chesbro—and in the minds of fans—for years. According to *Baseball's Best: the Hall of Fame Gallery* by Martin Appel and Burt Goldblatt (McGraw-Hill, 1980), "The disappointment of that season stayed with Chesbro for a long time. He was a successful lumber salesman in North Adams during the off seasons, and customers would always ask about the wild pitch, forgetting his 41 victories."

Later Chesbro's widow tried to have the wild pitch blamed on the catcher as a passed ball. But Chesbro never blamed anyone but himself for the lapse.

173

WHY THE PITCHING DISTANCE IS 60′ 6″

Many sports announcers like to refer to baseball as "a game of inches." But the most surprising inches in the sport may well be the six inches in the official distance of 60 feet 6 inches from pitcher to home plate. Why the six inches?

A mistake, of course.

In 1893 baseball officials decided to change the pitching distance from 50 feet to 60 feet. The diagram showing this read 60′ 0″. The surveyor who followed the hand-written instructions in mapping out the field mistook the 0 for a 6. The extra 6 inches has stayed ever since in the official pitching distance in professional baseball.

THE BRANCH RICKEY RECORD

One of the greatest names in sports history is that of Branch Rickey. In 1947, as president and general manager of the Brooklyn Dodgers, he put on the team the black baseball player Jackie Robinson, thereby breaking the color barrier in major-league baseball.

Rickey holds another honor, this one of dubious distinction. He himself played in 119 major-league games early this century—as outfielder, as first baseman, and, for 66 games, as catcher. But he was obviously a blunderhead as a catcher. In one game in 1907, he had 13 bases stolen off him—a modern baseball record still standing more than seventy-five years later.

WHEN LOU GEHRIG'S BASERUNNING COST HIM HIS OWN HOME RUN TITLE

New York Yankee great Lou Gehrig (1903–1941) always seemed to play in the shadow of Babe Ruth. Ruth, wearing No. 3, batted in the spot before Gehrig, wearing No. 4. Gehrig had a whopping 493 lifetime home runs and .340 career batting average, but the Babe hit even more and captured numerous home run crowns. In all the years they were teammates, Gehrig was able to achieve only one home run crown to himself—in 1934, when he hit 49.

Ironically, a mental error kept Gehrig from beating out Ruth for another home run title. In 1931 Gehrig and Ruth finished the season tied at 46 home runs apiece—but this was only because the Yankee first baseman lost his forty-seventh due to a base-running mistake.

On April 26, 1931, with a man on first and two out, Gehrig smashed a home run into the center-field stands. The ball hit a seat and bounced back onto the field where a fielder grabbed it, but the base runner, Lyn Lary, thinking the ball had been caught for the third out, stopped running and walked off the field. Gehrig, running with his head down, did not see what Lary was doing and continued to run the bases. When he reached home plate, the umpire called him out for passing a base runner, and Gehrig was credited with only a triple on his homer into the center-field stands.

As a result, at the end of the season, Lou Gehrig wound up having to share the home run crown with Ruth because of his blunder on the base path.*

THE HOME RUN KING WHO LOST A HOME RUN

Hank Aaron hit more home runs than any other player in baseball history. He retired with 755 career home runs, 40 more than fabled Babe Ruth. But on his first home run in professional baseball, Aaron committed an error so elementary that he was called out and lost the four-bagger.

It was in 1952, while he was playing Class C ball in Eau Claire, Wisconsin, that Aaron slammed his first homer as a pro. He became so excited that as he began to run the bases, he missed first base. The opposing team appealed, and Aaron was called out.

But Aaron learned from his blunder. According to an article in the *New York Times* on his career, he thereafter never watched the flight of his many home runs but concentrated on touching first base.

THE MISTAKEN BATTING TITLE

Ty Cobb, the famed Detroit Tiger outfielder, holds the batting title for the American League for 1910 with a .385 average. However, recent

---

*Gehrig also made a glaring flub off the field. Appearing in a paid radio commercial for a cereal, he was asked by the announcer how he had developed into such a great athlete. Gehrig responded, "Eating Wheaties." There was just one problem. The sponsor was Huskies, a competing brand to Wheaties.

checking of that season shows that a game in which Cobb went two for three was erroneously counted twice. Nap Lajoie, who hit .384 that year, should really have been named the winner of the batting title. The record books, however, still incorrectly list Cobb as batting champion for 1910.

SPECIAL MILESTONES IN THE MARCH OF BASEBALL ERROR

In the 1920 World Series between the Brooklyn Dodgers and the Cleveland Indians, Dodger pitcher Clarence Mitchell hit into a triple play, then followed that his next time at bat by hitting into a double play.

In the midst of his famous batting streak of hitting successfully in 56 straight games during the 1941 season, Joe DiMaggio also experienced one of baseball's worst outbreaks of fielding foul-up. He committed 4 errors in one day. During a double header on May 30, 1941, against the Boston Red Sox, the New York Yankee outfielder misplayed a ground ball, dropped a fly ball, and twice, while trying to throw runners out at the plate, heaved the ball all the way into the grandstand.

In 1925 Roger Peckinpaugh, shortstop for the Washington Senators, made only 28 errors during the entire season—but committed 8 errors during the World Series.

The worst fielding team in history was the Detroit Tigers of 1901. They committed 425 errors, averaging nearly 3 a game. In contrast, the team with the fewest errors was the Baltimore Orioles of 1964 and 1980 with 95.

The record for most errors in one game is held by Andy Leonard of the Boston Braves. His achievement: 9 errors in an extra-inning game.

Dick Stuart, who played for the Boston Red Sox, Pittsburgh Pirates, and Philadelphia Phillies, led the major leagues in errors for a first baseman seven times. A Boston sportswriter labeled him Dr. Strangeglove.

* * *

Third baseman Brooks Robinson and shortstop Mark Belanger played together on the Baltimore Orioles, where they were renowned for their fielding ability (they each won numerous Gold Glove Awards). But they both achieved a dubious distinction: They committed 3 errors in one inning. In fact, Belanger made his 3 errors on three successive batters.

Many players, too numerous to mention, share the record for 4 errors in one inning and 5 errors in a nine-inning game.

A PLAYER WHO KNOWS HIS PLACE

Graig Nettles, former New York Yankee third baseman now with the San Diego Padres, has won numerous Gold Glove Awards for his fielding ability. But Nettles is not above some self-deprecating humor. Recognizing the fielding mistakes every baseball player must live with, he has on the back of his glove the notation "E-5." This is the baseball scorer's symbol for "Error by the Third Baseman."

PRACTICE DOESN'T NECESSARILY MAKE PERFECT

How difficult is it for a pitcher not to make a mistake? It may be one of the most difficult things in sports.

During the course of the tens of thousands of major-league games played since records have been kept, only eleven perfect games have been pitched. A perfect game is defined as one in which a pitcher does not allow any batter to get on base or to score a run. Although in a no-hitter a batter could reach base through a walk or error, in a perfect game the pitcher faces the minimum number of opposing batters— twenty-seven—getting every batter to strike, fly, or ground out. It means that not only the pitcher but the fielders behind him do not make a mistake for nine innings.

Even more difficult a feat may be the pitching of a perfect game in a World Series. Only one such game has occurred. It happened on October 8, 1956, during the fifth game of the World Series between the New York Yankees and Brooklyn Dodgers. Don Larsen, a pitcher for the New York Yankees, who at the time had actually lost ten more games than he won over four previous seasons (two years before, he was 3-21, and his overall record was a paltry 30-40), beat the Brooklyn Dodgers, 2–0, not letting a single Dodger get to first base.

Following the historic game, a reporter asked Don Larsen, "Is that the best game you ever pitched?"

THE EIGHTH WONDER WITH 4,596 FAULTS

Errors occur not only on the field but over the field.

On April 19, 1965, the first domed all-weather stadium opened—the Houston Astrodome, built then for a princely sum of $30 million. Heralded as the world's largest air-conditioned room and a facility that would change playing conditions for sports, the Astrodome was glibly called the Eighth Wonder of the World.

The only problem was that the Eighth Wonder had 4,596 faults. This was the number of transparent plastic panels in the dome through which sunlight streamed, creating a glare for players looking up at the ceiling. In addition, the gray steel girders of the dome, combined with the panels, formed a crazy-quilt background to any fly ball.

The mistake in the design of the dome became readily apparent during the first baseball game played in the Astrodome with the sun shining outside. In a contest between the Houston Astros and the Baltimore Orioles, there were six errors, three of which were due to the problem with the dome, with players almost committing three other dome-related miscues.

More than 1,000 solutions to the sunshine problem were offered by fans, engineers, and baseball executives, including the use of special sunglasses and orange baseballs. Finally, the dome's skylight difficulty was resolved when the panels were painted a dark color to shut out the sun. But the $30 million first attempt to enable an outdoor sport to be played indoors had started with a glaring mistake.

THE SCOREBOARD TYPO

Typos can be embarrassing, but what about the typo that is in giant letters and is flashed before a stadium of people? This occurred during a game of the Oakland Athletics in the Oakland Coliseum, when its giant message board lit up with a one-letter flub on a question about baseball. The intention was to ask the crowd who was the record holder for number of bases in a baseball season. Instead, fans had to ponder this query:

"Who holds the record for the most babes in a single season—Hornsby, Musial, Ruth, Cobb?"

Our guess is that the answer is Ruth . . . to both questions.

WHEN THE MANAGER'S CLICHÉ RICOCHETED

Baseball managers always like to look for the silver lining in a situation, especially when the only thing left is a silver lining. One favorite cliché of a manager seems to be the statement "It's not over till it's over," meaning that a team is not out of a pennant race until it is mathematically eliminated, no matter how far back the team might be or how badly it is playing.

One time, however, Danny Ozark, then the manager of the Philadelphia Phillies, went too far in using the cliché. His team had just lost a game and had been mathematically eliminated, but when baseball writers asked him after the game, "What do you think, Danny?" he dutifully replied, "It isn't over till it's over."

This time one of the writers had to tell the manager, "It's already over, Danny."

BERRA BON MOTS

Yogi Berra, the New York Yankee catcher who is in the Baseball Hall of Fame (he was voted the American League's Most Valuable Player three times), was known for his unusual way with words. At the ceremony inducting him into the Hall of Fame, he began his speech by saying, "I thank everybody for making this day necessary."

Although his sentences often came out incorrectly, they seemed to contain the elements of truth and were, beneath the surface, quite understandable. For instance, Berra once remarked, "If people don't want to come to the ballpark, how are you gonna stop them?"

Berra's most cogent statement, especially as it relates to people's propensity to commit errors, came after a lost ball game. His unusually worded declaration summed up the athlete's battle with bungling. Said Berra: "We made too many wrong mistakes."*

A CASE OF STENGELESE

Casey Stengel was one of the most successful managers in baseball history. At one point his New York Yankee teams won five straight championships—a feat still to be matched. Yet Stengel butchered the

---

*Interestingly, while Berra was known to bobble language off the field, he rarely bobbled on the field. One of his records: Between July 28, 1957, and May 10, 1959, he caught 148 games without committing an error.

English language worse than his catcher Yogi Berra. While Berra at least could be understood beneath the verbiage, Stengel talked in a fractured double talk that even an interpreter of diplomatic speeches at the UN would find difficult to unravel. Here is just one example of Stengelese:

> No manager is ever gonna run a tail-end club and be popular because there is no strikeout king that he's gonna go up and shake hands with and they're gonna love ya because who's gonna kiss a player when he strikes out and I got a shortstop which I don't think I coulda been a success without him if ya mix up the infield ya can't have teamwork and it's a strange thing if ya look it up that the Milwaukee club in the morning paper lost a doubleheader and they got three of my players on their team and you can think it over.

The amazing thing is that the man was ever able to get hired as manager, let alone communicate with his players.

THROWING OUT THE FIRST BALL

One of the great traditions in baseball is the throwing out of the first ball to start a World Series game. Many times the one honored with the first ball is the president of the United States. The first chief executive to open a World Series was President William Howard Taft in 1910.

In 1931 President Herbert Hoover, who was flubbing the economy, managed to flub this simple act, too. With the country mired in the Depression, President Hoover was eager to throw out the first ball of a World Series game because he felt it would boost the nation's spirits to see him openly involved in a great sports event. On the day of the game he traveled by special train from Washington, D.C., to Philadelphia, where the World Series was being played between the St. Louis Cardinals of the National League and the Philadelphia Athletics of the American League.

Upon arrival at the stadium Hoover was besieged by news photographers, who asked him to stage throwing out the first ball before the game. He obliged the many requests for so long that baseball officials, seeing him preoccupied, went ahead and started the game without him.

Later that day a newsman, trying to remember if Hoover had officially thrown out the ball as part of the opening ceremonies, went to check. He found that Hoover still had the ball. After posing for so many photographs, the president, seeing that the game had started, had simply put the baseball into his pocket and sat down. Although he had made a special effort to throw out the first ball, President Hoover had missed his opportunity to open a World Series game.

OUR "ABNER DOUBLEDAY AWARD" FOR MOST EMBARRASSING ERROR COMMITTED BY A COMMISSIONER OF BASEBALL

One of the most embarrassing errors in baseball history was committed off the field by the commissioner of baseball during the 1981 season.

At the conclusion of a prolonged players' strike that had stopped the games in midseason, Commissioner Bowie Kuhn, whose salary ran into the hundreds of thousands of dollars annually, had to determine how the season should resume and how teams should be selected for the World Series. After much thought he presented a plan which called for splitting the season into two segments, with won and lost records and division standings tabulated separately for the two parts.

At the end of the season a play-off was to be held within each of the divisions before the winning team could proceed, as before, into a league championship and then on to the World Series. The key provision in the plan for determining who qualified for the play-offs was that if the leader in the first half of the season also came out on top in the second half, then the team with the second best percentage for the entire season would become eligible to appear in the play-offs.

The convoluted play-off system was seen by critics as unnecessary and a ploy to set up another round of lucrative play-off games to enrich baseball coffers. Many felt that a team that had the best record over both halves was the rightful division champ.

Kuhn resisted such criticism. Then a fan, followed by the manager of the Chicago White Sox, Tony LaRussa, pointed out that this system had a major defect which touched on the very heart of the game: A team could find itself actually benefiting by purposely losing certain games. Under the right circumstances a team with the second best record but no chance to come in first could secure a spot in the play-offs by dropping games that would help determine that the winner of the first segment would win again in the second segment rather than another team. With the same team coming in first twice, this would mean that the team with the second best percentage throughout the season would now enter the play-offs—even though that team had never won either half of the season.

Sportswriters seized on the mistake; fans complained; players either chortled or grumbled. A ground swell of opposition to the plan surfaced. The commissioner was urged to change the system. After much prodding Kuhn finally acknowledged the error, and the owners revised the much ballyhooed plan.

The media and the fans were only partially satisfied. They hooted the baseball leadership for the fumbling way in which the aftermath of the strike had been handled and for trying to milk the situation for more money. The *New York Times* in an August 22, 1981, editorial, "Winning Is Not the Only Thing," noted how "the lords of baseball" had been "embarrassed by a clubhouse lawyer."

A columnist for a newspaper in the Los Angeles area also criticized the commissioner and suggested that Bowie Kuhn be replaced with Los Angeles Dodger owner Walter O'Malley since "he's been running baseball for years, anyway."

There was only one problem with this idea: O'Malley had been dead for years.

When Baltimore Oriole owner Edward Bennett Williams, no supporter of Kuhn's, was advised of the writer's suggestion to have the deceased O'Malley replace the commissioner, Williams remarked, "You can't say we didn't have a precedent. We've already got a dead man filling the job."

A year later baseball owners voted not to renew Kuhn's contract. It shows that in the world of sports it may be human to err, but there is little forgiveness in Mudville. Error puts players into the record books and commissioners into the unemployment line.

WHY UMPIRES ARE SAID TO BE BLIND

Leave it to an umpire to make the most sweeping, asinine assertion about personal infallibility in the face of widespread fallibility.

Bill Klem, who officiated in a record eighteen World Series spanning 1908 to 1940, once declared, "I never made a wrong call in my life."

On the Football Field

THE FOOTBALL GAME CHANGED
BY A REFEREE'S ERROR

Many sports fans have often wondered if a particularly close or contested call by a referee or an umpire could ever be reversed after a game. Rarely, if ever, has this occurred, but one time a referee later not only publicly admitted that he had made an error but changed the score of the game and its winner because of his mistake.

On October 21, 1922, the Columbia and New York University football teams were locked in a close game before a large crowd. In the first quarter a Columbia player, deep in his own territory, tried to punt, but an NYU player burst through the line and blocked the football, which bounced back over the goal line and into the end zone. Another NYU player ran after the ball, trying to stop it before it went back through the end zone, but could get to the ball only after it had bounced past the end line. Nevertheless, as soon as the NYU player gained control, the referee, William N. Morice, ruled a touchdown for NYU.

In the final quarter of the game, with the score still 7–0 in NYU's favor, Columbia finally scored, but its point after touchdown was blocked. The 7–6 score in favor of NYU held through the final minutes of the game. NYU fans went home elated, believing their team had won.

But as Referee Morice went home, he began to have doubts about his call on that first scoring play. Some people were saying he had made a mistake and, instead of ruling the play a touchdown, should have ruled it a safety because the ball had bounced out of the end zone before NYU could gain control. If so, then NYU should have got only 2 points for a safety on the play. However, then Columbia would be the victor with a score of 6–2 instead of a loser to NYU with a score of 7–6.

Morice came to the same conclusion. Incredibly he reversed himself and announced Columbia was really the winner of the game with a score of 6–2 instead of losing by 7–6. Morice announced: "In justice to Columbia, I feel that I must publicly admit my error and reverse my decision on the play in question, so that the final official score should have been 6–2 in favor of Columbia. I wish to express my very great regret to both teams, to Columbia for having deprived them of a victory

183

at the time, and to New York University for having to reverse my decision at this late date. . . . I feel that I cannot do otherwise."

Morice's expression of error and change of ruling caused a great debate to erupt. A number of football officials termed Morice's admission of error "the right and manly thing." But others pointed out that once Morice had made a ruling during a game, especially one that directly affected the score, the score should hold since that was now the reality of the game to which both teams had to respond. Some questioned the power of a referee to change a score after a game had concluded.

The controversy over that game has never really ended. Columbia lists the game in its record as a 6–2 win over NYU, while NYU, ignoring Morice's change of heart, records the contest as a 7–6 victory over Columbia. In the general record books on college football, the game is listed with an asterisk, a fitting sentinel of a referee's mistake and his uncommon act of publicly recognizing his error.

THE TIME WHEN REFEREES PAID FOR THEIR MISTAKE

Players are often fined for infractions on the field. But once the referees in a game were fined for making a mistake that could have affected the outcome of the contest.

On December 8, 1968, the Los Angeles Rams and the Chicago Bears were locked in a key National Football League Game (the Rams were in a tight race with the Baltimore Colts for the division championship). Behind by 17–16 and with only seconds left, the Rams advanced toward the Bear goal line, hoping to get within field goal range. On first down, the Ram quarterback, Roman Gabriel, threw an incomplete pass, but because a Ram player was caught holding, the Rams were penalized back to the 47-yard line of the Bears. On the next three plays, Gabriel failed to complete a pass. With ten seconds remaining, referee Norm Schachter, along with his umpiring crew, turned the ball over to the Bears, who ran out the clock. But overlooked by the referees was a missed down for the Rams: The holding penalty assessed the Rams at the beginning of their last series of downs meant the Rams still had one more down coming to them.

When this lapse was discovered after the game, the National Football League fined the referees $250 to $1,650 for their mistake. But the score was allowed to stand, and the Rams still lost, 17–16.

FUMBLE CHAMPS

The football player who has fumbled more times than any other in football history is Roman Gabriel. He was a quarterback for the Los Angeles Rams (1962–1972) and the Philadelphia Eagles (1973–1977). His record performance: 105 fumbles.

The record for most fumbles in a season is held by another quarterback, Dan Pastorini. Playing with the Houston Oilers in 1973, he saw the football squirt out of his grasp 17 times.

Two teams are tied for the most fumbling teams in a season. The Chicago Bears of 1938 fumbled 56 times—a record that stood for forty years until the San Francisco 49ers of 1978 tied the record with 56 of their own fumbles.

WHEN A MISTAKE NULLIFIED A RECORD

January 10, 1983. On a football field in Cincinnati, one of the great running games of all time is occurring. Freeman McNeil, fullback for the New York Jets, is giving such an electrifying performance in a vital play-off game against the Cincinnati Bengals that as the *New York Times* later reports, "even newsmen and scouts in the Cincinnati press box whistled at some of his moves."

Indeed, McNeil that day has two runs of 35 yards each and one 20-yard gallop for a touchdown and has even thrown a touchdown pass. His runs are such remarkable, twisting affairs in which McNeil changes directions without appearing to slow down that another player, later reviewing films of the game, remarks, "I'd need four operations if I cut like that."

With four minutes remaining in the Jets' 44–17 victory, McNeil reaches what the official statistician for the game says is a staggering 206 yards rushing—which would tie him with the previous running record for a National Football League play-off game held by Keith Lincoln of the San Diego Chargers in 1963. But with the game safely won, McNeil has been removed from the game. Two publicity men with the New York Jets hurry to the sidelines to tell coach Walt Michaels that McNeil has a chance to break the rushing record. Since the game is virtually won, the New York coach tells his offensive coordinator, Joe Walton, "OK, let him go in for one play and then take him out."

McNeil is sent back into the game and rushes for another 5 yards, which everyone now believes gives him 211 yards for the day. He comes off the field to the congratulations of his teammates for his record-breaking achievement. Not until after the game, however, is a mistake uncovered in the statistics kept that day.

A teammate of McNeil's is the first to spot something amiss. Reading the statistics sheet after the game, tailback Bruce Harper notices that he is given credit for zero yards on one carry. But shortly before the first half ended, Harper ran for a 9-yard gain. He mentions the discrepancy to McNeil, but both think little more about it.

Later a Cincinnati newsman notices the same mistake: that Harper's run is missing on the statistics sheet. The next morning the reporter calls the league office to inquire about the missing play. The league in turn calls the Jets, and the offensive coordinator, Joe Walton, reviews the play-by-play sheet. He confirms that Harper had made a 9-yard gain, and it is missing on the sheet because it has mistakenly been entered as McNeil's.

Thus, as the game progressed throughout the second half, McNeil's rushing total, as tabulated by the statistician, who was an employee of the Bengals, was 9 yards more than he had actually gained. The New York Jets removed him from the game, thinking he had beaten the record by 6 yards, when in actuality McNeil had amassed just 202 yards —5 yards short of breaking a record that had stood for twenty years.

The error was never fully explained, but it was noted that McNeil's uniform number is 24, Harper's the reverse of that, 42.

After the official scorer's mistake was uncovered and the correction made, a subdued McNeil was quoted as saying, "Winning is all that matters." But the error that led to his being pulled from the game 5 yards too soon meant that what could have entered the record books as the greatest rushing performance ever in the National Football League play-offs was for Freeman McNeil just another hard day's work.

THE BEST OF ALL POSSIBLE BETS

Bettors are always looking for as much help as they can get. In the middle of December 1982 numerous bettors on football games must have had to rub their eyes in disbelief when they looked over the *New York Daily News* "Latest Line" column, in which the point spreads and betting favorites are listed. The *Daily News* printed the following point spread for an upcoming Raiders-Rams game:

FAVORITE	PTS.	UNDERDOG
LOS ANGELES	7½	LOS ANGELES

TV SPORTS' MOST EMBARRASSING MINUTE

Broadcasting coverage has so invaded the sports world that the mistakes of television executives can affect sports fans as much as the errors of athletes can.

Probably the most notorious example of this occurred on November 17, 1968, when the New York Jets played the Oakland Raiders in a nationally televised game on NBC. The Jets were leading, 32–29, with the clock showing 1:05 left in the game but the time in the real world at 6:59 P.M. EST.

Someone at NBC then made the decision to stop coverage of the game and shift to the network's regularly scheduled 7:00 P.M. Sunday movie, which that night was *Heidi.* Thousands of irate viewers began calling their local stations when their screens suddenly went from showing a football game involving sweating, struggling behemoths to a movie about a sweet girl living in the Alps. To add rage to the insult, the last minute of the now blacked-out game saw the Oakland Raiders stage a dramatic comeback and win, 43–32, with literally two last-minute touchdowns.

The anger of the fans now knew no bounds, and NBC and the nameless TV executive who gave the orders to switch coverage were the object of sporting world scorn. That game henceforth came to be known, according to *Sports Illustrated,* as "The Heidi Game—a memorial to the ability of network television to foul up."

So long has that memory lingered that fourteen years later *Sports Illustrated* used the name of the offending movie in an article on the worst examples of TV coverage of that year's sporting events. The magazine gave the dubious winners "our Heidi Awards for 1982."

THE PLAYER WHO SCORED FOR THE OTHER TEAM

Defensive end Jim Marshall of the Minnesota Vikings, who holds the professional football record for playing the most games in a row (282), also holds the rare and dubious distinction of giving the opposing team points. He did this early in his career on October 28, 1964, during a tense game with the San Francisco 49ers. When he saw the other team fumble, he scooped up the loose football, and, forgetting where his team's goal was, ran 60 yards the wrong way into his own end zone— for a safety and 2 points for the other side. Not until a player on the 49ers came over to him in the end zone and thanked him did he realize his blunder. Minnesota, however, won, 27–22.

THE ROSE BOWL AND WRONG WAY RIEGELS

Jim Marshall's wrong way run actually had a precedent—this one in a major college football game. On January 1, 1929, in a Rose Bowl game between University of California and Georgia Tech, a California player scooped up a fumble and ran 60 yards the wrong way. He was stopped from scoring a safety for the other team only when a teammate virtually tackled him on the 2-yard line. But the blunder soon led to a score— and a lost game.

The play took place during the second period of a scoreless game. With the ball on his own 26-yard line, a Georgia Tech ball carrier raced through the line for 10 yards, but then was hit and fumbled. With the ball bouncing around on the forty-yard line, Roy Riegels, center and captain-elect for the University of California, grabbed the football in the air.

Riegels actually started running the right way—for about 10 yards. Then, inexplicably, he turned and started running toward his own goal. One of his teammates, Benny Lom, seeing Riegels' mistake, began running after him, but the pursuit only seemed to spur Riegels to run faster. Finally, at the California 10-yard line, Lom caught Riegels, grabbed him in a bear hug and tried to turn him around to run back up the field. Lom's efforts stopped Riegels from crossing into the end zone, but before Riegels could start running the right way, Georgia Tech players tackled him on the 1-yard line. Riegels, stunned, sat on the ground in disbelief as his teammates came over to console him.

Now in poor field position, California decided to punt, but the ball was blocked, touched a California player, and went back through the end zone for a 2-point safety for Georgia Tech.

Later, Georgia Tech scored a touchdown for another 6 points (they missed the extra point) and then hung on for an 8–7 victory.

Ever since, Roy Riegels has been known as the player who lost the Rose Bowl by running the wrong way. "Even now," he was quoted years later, "you see the words 'wrong way' and you think they're referring to you. . . . I still don't understand how I did it wrong."

At the Olympics

THE FIRST NUDE OLYMPICS

The modern Olympics date from 1896, when the ancient idea of athletic competition between nations was revived in Athens. But today's Olympics differ from the ancient Greek games in one important way: Many of the original Olympics were conducted in the nude. And the naked truth is that such nudity resulted from an accident.

During the fifteenth Olympics in 720 B.C., while running in a footrace, an entrant—his name was Orsippus of Megara—suddenly lost his loincloth. However, in spite of losing his clothing, he won the race. A remarkable reaction then set in among other runners. They ascribed Orsippus' victory to the absence of clothing, and soon many of them chose to run in the nude. Their reasoning was certainly correct. The absence of clothing reduced wind friction and enabled runners to reduce their running times. Soon other athletes, liking the freedom of movement, threw off their loincloths, too. Thus, the ancient Olympics became a nude event, thanks to the accidental fall of a loincloth.

HOW THE GREAT ZAMORA ERRED

In the 1924 Olympics Ricardo Zamora of Spain was considered the soccer world's greatest goalkeeper. Dubbed the Great Zamora, he boldly proclaimed he would not allow the opposing team to score a single goal against his team in the soccer championship.

In the very first game of the competition, the Great Zamora seemed to be keeping his word as he turned back all shots by opposing Italian players. But then, in the last minute of play, with no score, the captain of his own team inadvertently kicked the ball back toward Zamora. To everyone's astonishment, including Zamora's, the ball went past Zamora and into his net, scoring the only point of the game for Italy and knocking Zamora's team out of the competition. The Great Zamora fell to the ground and cried openly.

IF LOOKS COULD KILL . . .

During the 1932 Olympic Games in Los Angeles French discus thrower Jules Noel produced a shot that many observers considered the longest of the event. However, just as he was throwing the discus, the field judges suddenly were distracted by a pole vault elsewhere on the field and forgot to watch Noel. Since none of the judges had seen where his throw had landed, they made him take his throw over. But Noel could not equal his original throw, nor could he now come close to the leaders in the competition. In the final listing of the event Jules Noel is fourth.

THE SADDEST OLYMPIC MISTAKE EVER MADE

To Wim Esajas belongs possibly the saddest mistake ever made in the Olympics.

Esajas, a competitor in the 800 meters, was the one and only representative from Surinam, Dutch Guiana, at the 1960 Olympics in Rome. As the day of his event came, Esajas decided to rest at the Olympic Village before going to the track for the afternoon competition.

There was only one problem. When Esajas arrived in the afternoon, he discovered that the 800-meter heats had been run in the morning, and he had been eliminated for not showing up. Wim Esajas had to return to Surinam without ever competing.

THE EXCEPTION THAT PROVES THE RULE

The difficulty in achieving an error-free performance can nowhere be better seen than by looking at how rare was the performance of a fourteen-year-old girl in the 1976 Olympics in Montreal.

Nadia Comaneci was a four-foot-eleven-inch, eighty-six-pound gymnast from Rumania who was competing in her first Olympic Games. She recorded a perfect score of 10 points on the balance bar, beam events, and the uneven bars. In the uneven bars, for instance, she earned an overall score of an amazingly perfect 20.00. She went on to win gold medals in uneven bars, beam, and best all-around.

Nadia Comaneci thus became the first individual to receive a perfect score of 10 in gymnastics in over 80 years of Olympic competition. (Men have been competing in Olympic gymnastics since the inception of the modern games in 1896. Women gymnasts began competing in the Olympics in 1928.) Since then, a number of gymnasts have received perfect scores, causing Olympics officials to think about changing the scoring system.

THEY SAID IT . . .

Four out of five points are won on your opponent's errors. So just hit the ball back over the net.

—Billy Talbert

Golf is the only sport I know of in which a player pays for every mistake. A man can muff a serve in tennis, miss a strike in baseball, or throw an incomplete pass in football and still have another chance to square himself. But in golf every swing counts against you.

—Lloyd Mangrum

Golf is the only game where the worst player gets the best of it. . . . The good player gets worried over the slightest mistake, whereas the poor player makes too many mistakes to worry over them.

—David Lloyd George

. . . AND HE SAID IT

I love sports. Whenever I can I always watch the Detroit Tigers on radio.

—President Gerald Ford

In Other Sports

GREETINGS FOR THE FIRST
FEMALE CHANNEL SWIMMER

Gertrude Ederle, a teenager from New York City, was the first woman to swim the English Channel. She accomplished her feat on August 6, 1926, swimming from the coast of France to Kingsdown on the English coast in fourteen hours and thirty-one minutes. Her time was nearly two hours faster than that of the previous fastest swimmer of the

Channel—male athlete Enrique Tiraboschi, who had performed the feat in 1923. Ederle thus showed not only that a woman could swim the English Channel but that she could do it faster than a man.

The day on which she accomplished this, however, was also the day on which the *London Daily News* editorialized that "even the most uncompromising champion of the rights and capacities of women must admit that in contests of physical skill, speed, and endurance they must remain forever the weaker sex."

LOST AT THE RACES

Eddie Arcaro was one of the most successful jockeys of all time. In thirty-one years of racing, he rode 4,779 winners, placing him high atop the all-time winning jockey list (he was the first jockey to win racing's Triple Crown twice). However, Eddie had some difficulty learning his trade and avoiding the errors apprentice jockeys so frequently make. He suffered 45 straight losses before guiding a horse to victory.

TYPO WINS PREAKNESS

In 1983 the winner of the 108th running of the Preakness Stakes—the second leg of the Triple Crown in horse racing—was Deputed Testamony. There is no such word as *testamony*. The word is *testimony*. The misspelling occurred when the horse was named. The owners never got around to making the correction.

WHEN THE HUNTER WAS THE HUNTED

In New York State in 1977 hunters were officially recorded as killing 83,204 deer, 1,369 wild turkeys, 551 black bears, and 7 fellow hunters.

HOW EISENHOWER GOOFED AT GOLF

President Dwight David Eisenhower was, as everyone knows, a great golf enthusiast. Almost every Sunday he played golf at a course near his home in Gettysburg, Pennsylvania. What Eisenhower and most people did not know, however, was that until 1960, when it was repealed, a 1794 state law made it illegal to play golf in Pennsylvania on Sunday.

BLOOPERS IN THE BOOTH

The tension and speed involved in most athletic events take a heavy toll of the lips and tongues of sports announcers. As a result, bloopers abound in the broadcast booth. Some examples, culled from George Gipe's *The Great American Sports Book* (Doubleday, 1978):

—The all-star football game dragged on through a constant downpour, leading Curt Gowdy, TV network announcer, to comment on how flooded the field was and to remark, "If there's a pileup, they'll have to give some of the players artificial insemination."
  Some other Gowdyisms:

—"Brooks Robinson is not a fast man, but his arms and legs move very quickly."

—"Willie Mays won't start today because the Mets' regular outfielder Rusty Slob will be playing. Wait, that's Rusty Staub!"

—"And at the game's end, it's National League 6, American League 4. That score again is American League 6, National League 4."

—"Luis Tiant comes from everywhere except between his legs."

—"To give you a better idea of how the teams shape up against each other, we'll be throwing up statistics like these all during the game."

—"Folks, this is perfect weather for today's game. Not a breath of air."

But Gowdy is not alone. Consider:

—A Canadian sportscaster commenting on a hockey game: "Toronto hockey fans will be glad to learn that their goalie made his first girl ever in the last 10 seconds of play."

—An announcer at a golf tournament: "Arnie Palmer is getting ready to putt. Arnie, usually a great putter, seems to be having trouble with his long putt. However, he has no trouble dropping his shorts."

—Another broadcaster of a golfing match: "Johnny Tee is now on the pot . . . er, I mean Johnny Pott."

—Jerry Coleman, radio announcer of New York Yankee games: "There's a fly ball deep to center field. Winfield is going back, back . . . he hits his head against the wall. It's rolling towards second base."

—The announcer of sports news in Austin, Texas: "And now for the gay dames—no, uh, the day's games."

Among Spectators

Sports, especially professional sports, exist in large part for the entertainment of fans and spectators. We conclude our look at error in sports with some incidents in which mistakes led to injury not for the athlete but for the spectator.

WHY BOXING IS DANGEROUS

During the second Jack Dempsey-Gene Tunney bout (September 22, 1927), one of the most dramatic boxing matches in history, a listener in Los Angeles became so excited while cheering that he stabbed himself with an ice pick.

HOLE IN ONE

There have been many instances in which a golfer misdirected a shot and hit a bystander. As president, a golf-playing Gerald Ford hit several people this way. So did Spiro Agnew when he was vice president. In fact, one day at the golf course (February 13, 1971), Agnew hit a husband and wife with just one shot—and then on his very next swing hit a woman on the ankle. He had had good practice for this. The previous year Agnew had struck golf pro Doug Sanders in the head.

But no golfer may have predicted an errant shot as well as did pro golfer Mac McLendon.

After doing poorly on the first round of the 1979 Masters Tournament, McLendon remarked to his wife, Joan, "I just know I'm going to hit someone."

The next day, on his very first swing at the first hole, McLendon drove his tee shot into the crowd and hit his wife, Joan.

SPECTATOR COMFORT

On February 17, 1974, one of the largest losses of life involving spectators at a sporting event occurred in Cairo, Egypt—all because of a mistaken decision by promoters of a soccer game.

The soccer match between Egypt's National All-Star Team and a visiting team from Czechoslovakia was to be held in Cairo Stadium, which seats 100,000 spectators, but the contest was shifted to the Zamalek sporting club, which has a capacity of only 45,000. The problem was that 80,000 people showed up on the day of the game.

Attempts by fans to be the first to get into the smaller stadium caused first a stampede, next a riot, then a murderous crush of humanity. The result was 47 people injured, 49 dead.

After the disaster, promoters of the soccer match explained the switch of stadiums. It had been done, they said, "to ensure the comfort of the spectators."

WHEN A FAN IS ALSO A MOTHER

Some fans, it is said, live and die with their team's fortunes. But what happens when a fan is also a mother?

While the New York Giants lost one pennant through a player's mistake [see the story in this chapter of Fred Merkle and the boner that cost the Giants the 1908 pennant], just four years later the team suffered from the miscue of another player—and this time it cost them the World Championship . . . and almost a player's mother.

With the 1912 World Series between the Giants and the Boston Red Sox even at 3–3, the final and deciding game entered the tenth inning with the score tied 1–1. Fred Merkle, finally seeming about to earn a hero's halo, singled home what looked like the winning run.

But in the bottom of the tenth, the first Boston batter hit a soft fly ball to center field, a routine easy out. Giant center fielder Fred Snodgrass loped into position for the catch, put out his glove, and—incredibly—dropped the ball. The runner, who reached second base on the error, soon scored with the winning run of the game—and the World Series.

Poor Fred Snodgrass. His one miscue was headlined the next day in the *New York Tribune:* ERROR BY SNODGRASS THE DIFFERENCE BETWEEN WINNING AND LOSING A TITLE. Ever after, he was known by his misplaying of an easy fly ball.

But no one may have been as affected by the Snodgrass muff as was his mother, who was also known as an ardent baseball fan. A continent away at the time, Mrs. Snodgrass was in a Los Angeles theater following the game by means of an electrical scoreboard. When it showed that Fred had missed a fly ball and that her son's error had cost the New York Giants the World Championship, she did what any mother—and fan—would do. She collapsed.

The incident was reported in a separate news article. Its headline: FAINTS ON MUFFED FLY.

"YOU CAN'T BELIEVE EVERYTHING YOU READ IN THE PAPER" DEPARTMENT:

THE TIME WHEN DEWEY BEAT TRUMAN—THE ONLY TIME

It will take a long time for the journalism fraternity to live down the famous picture of a victorious Harry Truman holding up an edition of the *Chicago Daily Tribune* declaring, in World War II type, DEWEY DEFEATS TRUMAN. President Truman's broad smile was the heartfelt happiness of he who triumphs over someone else's mistake and has the last laugh. Many newspapers—and even public opinion polls—had been predicting a Truman defeat in the 1948 presidential election at the hands of the Republican, New York Governor Thomas E. Dewey. But Truman, who was finishing out his first term as president after the death of FDR, had conducted an aggressive campaign. The result was 303 electoral votes for Truman, only 189 for Dewey, plus a Democratic-controlled Congress, Democratic wins in many state and local contests —and numerous embarrassed newspapers and pollsters.

CHAPTER VII

Typo:
Error in the Library

*Errors of fact do more to undermine the trust
and confidence of readers than any other sin
we commit. A city editor I knew used to say:
"A story is only as good as the dumbest error
in it."*

—Donald D. Jones, ombudsman of the
Kansas City Star and Times,
in *Time* magazine Press article
"Why Readers Mistrust Newspapers"
May 9, 1983

Many forms of error do their damage and then evaporate
with the passage of time—the residue of their harm left only in a cloud
of dust and an expletive of disgust. Libraries, however, those reposito-
ries of study and scholarship, of careful research and rechecking, writing and rewriting, are really museums of error.

Libraries, you see, are filled with mistakes, with books that offer out
of date or incorrect information, with works that have preserved in type
the erroneous fact, the faulty research, the statements once believed by
scholars to be true that are now known by even the schoolchild to be
fiction.

Indeed, here under one roof is evidence of how our most studious minds, even filtered through editors, proofreaders, and publishers, are prone to mistake. We need only compare one respected volume with another to see how they differ in facts and statements, how inconsistencies or inaccuracies entangle our search for accuracy, for truth.

And what is the most frightening is that these errors nestle between book covers for generations, always on exhibit, always accessible to a passing parade of believing, trusting readers. The library is a reminder of how, more often than not, error creeps into our world and stays.

THE "GUTENBERG BIBLE"

Although long in existence, the library as we know it today, with inexpensively produced books widely available to the public, had its beginnings with the printing press. A figure we all know and admire for those early days of printing is Johann Gutenberg. Gutenberg (c. 1398–1468) is generally referred to as the inventor of movable type and printer of the first book from movable type. That book, called the Gutenberg Bible, is a Latin work of 1,282 pages now considered one of the rarest and most important printed books in history.

Our excursion into error in the library begins with Gutenberg because he was far from the important figure often cited. In fact, Gutenberg's business errors kept him from even being around when the Bible that bears his name was printed. Rather, there is ample evidence that the printing of that Bible owes as much if not more to a man named Johann Fust as it does to Gutenberg.

In 1450 Fust advanced Gutenberg 800 guilders to promote his printing efforts and to enable him to make the necessary tools to carry out his work. But Gutenberg committed so many blunders that by 1452 Fust had to advance another 800 guilders to him to prevent a collapse of Gutenberg's business. However, even this did not help, and by sometime before November 1455 Fust took legal proceedings against Gutenberg.

Under the original arrangement, Gutenberg was obligated to turn over all his printing materials and any works in preparation to Fust. One of those works was the famed *Bible of 42 lines,* as it was called (each column was prepared in groups of forty-two lines of type). Fust took this material to his own house in Mainz and there, with the assistance of Peter Schöffer, completed work on the Bible and published it to wide acclaim.

Thus, when the *Bible of 42 lines* was issued in 1456, Gutenberg was nowhere around. His involvement was limited to working on part of the Bible, but surely not all of it. The linking of his name with that Bible, as is now so commonly done, is far from correct. In fact, the *Encyclopaedia Britannica* points out that although "German bibliographers now claim this Bible for Gutenberg, . . . according to bibliographical rules, it must be ascribed to Peter Schöffer, perhaps in partnership with Fust."

Actually, once Gutenberg had his falling-out with Fust, Gutenberg seems to have disappeared from the world of printing. No books bearing his name or genuine portraits of him are known to exist. Many works today print a picture of a heavily bearded Gutenberg, but the *Encyclopaedia Britannica* states that all portraits of him "appearing upon medals, statues or engraved plates" are fictitious.

Gutenberg's fame is all the more surprising because he was not the first to use movable type or to print a book with it. The Chinese and Koreans, long before Gutenberg, were cutting text and pictures onto blocks of wood and printing from that. They had even developed movable types of porcelain and metal. The problem for them was one of language. The Chinese and Korean languages are so complex that they found this printing method impractical. The idea was not revived until Gutenberg's time.

What Gutenberg accomplished was to make printing from movable metallic type at least practical for the first time. His errors in business, however, removed him from the scene of one of the great advances in human history—the printing of what could well be said to be the Fust Bible.

THE FIRST PRINTED BOOK WITH ERRORS

Early books were written by hand, which led to many errors in transcribing and copying. Ancient manuscripts still in existence can be seen to have crossed-out words and reworked lines by scribes who caught a mistake and corrected it. The Dead Sea Scrolls, for instance, abound in such errors. But many times errors in those preprint scrolls and books went uncorrected, causing problems for later scribes, who did not know if an oddly spelled or used word, sentence, or paragraph was intended or not.

It was hoped that the introduction of movable type and the printing press would alleviate this problem. It did not. The first errors to creep into printed books occurred as early as 1478, in an edition of *Juvenal* printed in Venice.

THE WORST PRINTED BOOK

Nomination for the book with the most printing errors may well go to a little book issued in 1561. Entitled *Missae ac Missalis Anatomia,* it runs only 172 pages but contains 15 pages of errata. The author, trying to explain the march of misprints, declared that the Devil had made the printer do it: Satan, said the author, had soaked the manuscript to make it nearly illegible, thereby causing the printer to misread it and commit so many mistakes.

What is noteworthy about this fiasco is that the author did not blame the publisher, but blaming Satan was part of the erroneous views of the day about printing. It was felt this was a "black art" and the work of the Devil since church and secular authorities believed that exposing the common populace to books would lead them astray.

THE MOST EMBARRASSING BIBLE PRINTING ERROR: "THOU SHALT COMMIT ADULTERY"

In London in 1631 was printed an authorized edition of the Bible that caused a commotion. After it had come off the presses, this Bible was found to have no negative in the listing of the seventh of the Ten Commandments. It flatly told the reader, "Thou shalt commit adultery."

As a result of this error, the Bible came to be known as the Wicked Bible. Its printers, Robert Barker and Martin Lucas, were fined 3,000 pounds.

MORE BIBLE ERRORS

During the seventeenth century many Bibles were printed in cheap editions in Holland. As a result, errors crept in, and a large printing of Dutch English Bibles was burnt by order of the Assembly of Divines. One typo that really incensed authorities was a mix-up in a passage in

Ruth (4:13) in which the Lord, instead of giving her "conception" gave her "corruption."

Carlo Guidi (1650–1712), an Italian lyric poet, worked diligently translating a religious work into Latin. He was about to present the work to Pope Clement XI when he discovered a typographical error. The Latin word *sine* ("without") had been printed throughout as *sin.* Guidi took the mistake so seriously that he dropped dead of apoplexy at the discovery.

SHAKESPEARE'S SHAKY ACCURACY

William Shakespeare, who obtained many of his plots from history books and other sources, seemed to have had trouble getting and keeping all of his facts straight. The following are some errors in the works of Shakespeare:

> *In *Julius Caesar,* Shakespeare refers to a clock that strikes the hour—1,400 years before the invention of such a clock.
> *In *Hamlet,* the Ghost appears to be a Catholic (he speaks of purgatory and absolution), but during the time in which the play takes place in Denmark, the Danes were pagans.
> *Also in *Hamlet,* Shakespeare writes of Elsinore's "beetling cliff." Elsinore has no cliffs.
> *In *The Winter's Tale,* a vessel is said to be "driven by storm on the coast of Bohemia." Also, Antigonus remarks, "Our ship hath touched upon the deserts of Bohemia." The problem with both references is that Bohemia, now a part of Czechoslovakia, does not have a coast (the area is landlocked) and no ship could ever touch its deserts (the region is a fertile plateau surrounded by mountains and forests).
> *Delphi is a city, but in *Coriolanus,* Shakespeare calls it an island.
> *Shakespeare mentions billiards in *Antony and Cleopatra,* cannon in *King John,* and turkeys in *1 Henry IV*—all before their time of invention or discovery.

We should not be too hard on Shakespeare as a historian, though. In his will he spelled his own name four different ways.

ERRATIC ERRATA

Shakespeare is not the only author to make embarrassing mistakes in works that will endure for centuries with their errors present for all to see—also for centuries. Here are just some examples in the works of famous authors:

Miguel de Cervantes: In *Don Quixote,* Sancho Panza sells his donkey, then is seen riding it again (no explanation given). Sancho loses his wallet, only to be shown using it again later (no explanation given). Sancho loses his coat with food in the pocket; only later the food is shown to be still in his possession (no explanation given). Don Quixote's helmet is shattered into pieces, but later it is shown to be unharmed and in good condition (no explanation given).

Daniel Defoe: In *Robinson Crusoe,* Crusoe swims without clothes to a wrecked ship, finds some biscuits there, and then is said to put them into his pockets.

Sir Arthur Conan Doyle: In *A Study in Scarlet,* Sherlock Holmes's assistant, Dr. Watson, is said to have suffered a wartime bullet wound in the shoulder, but in *The Sign of the Four,* the wound is said to be in the leg.

John Keats: In "On First Looking into Chapman's Homer," Keats speaks of Cortez's discovering the Pacific Ocean. He didn't. Balboa did.

Sir Walter Scott: In *Ivanhoe,* one character has two different first names. At one point, Malvoisin's first name is Richard; another time it is Philip.

Leo Tolstoy: In *War and Peace,* Natasha is seventeen years old in 1805 and twenty-four years old in 1809— a growth of seven years in four. Prince Andrei's icon inexplicably changes during the novel from silver to gold.

Eugene O'Neill: In *Where the Cross Is Made,* the stage directions speak of a one-armed character sitting at a table "resting his elbows, his chin in his hands."

Vergil: In the *Aeneid,* Chorinaeus and Numa die, then later reappear without any reference to their deaths.

203

THE ATTEMPT AT AN ERROR-FREE BOOK

Once upon a time great care was taken by a publisher to produce a perfect book—free of any printer's errors. The publisher was the esteemed University Press in Glasgow, Scotland, and the work was the Foulises' editions of classical works, well regarded by nineteenth-century scholars and collectors.

Six experienced proofreaders were employed to check each page meticulously. After that the pages were put up in a university hall for two weeks, with the notification that 50 pounds would be given to anyone who could find an error on any of the pages. Only then, when the publishers believed every possible error had been discovered and corrected, was the work issued.

But upon publication it was discovered that the book contained several errors—including one in the first line of the first page.

WHAT ABOUT PERELMAN'S REVENGE?

Little in literary life can match an author's excitement at the publication of his or her first work. Consider, then, the birth of humorist S. J. Perelman's first book. A collection of forty-nine pieces entitled *Dawn Ginsbergh's Revenge,* it was brought out by Horace Liveright in 1929 in an edition that neglected one slight element. The title page lacked any mention of the author's name.

THE COOKBOOK THAT WAS
A REAL PUBLISHING BOMB

In 1978 Random House issued a cookbook that contained a potentially lethal mistake. *Woman's Day Crockery Cuisine* offered a recipe for caramel slices that inadvertently left out one simple ingredient—water. It was soon discovered that if the recipe were now followed, a can of condensed milk called for in the book could explode. Random House had to recall 10,000 copies of the book because of the hazardous lapse.

"ALL THE NEWS THAT'S FIT TO PRINT"

During the twentieth century the *New York Times* has developed into one of the leading newspapers of the world, viewed by librarians as a major resource and fount of information. But the august *New York Times* has erred, on numerous occasions.

One of the paper's most famous cases of mistake occurred on January 13, 1920, when the *Times* denigrated the rocket research of Robert H. Goddard and pooh-poohed the possibility of rocket travel through space. Editorialized the *Times:* "That Professor Goddard and his 'chair' in Clark College and the countenancing of the Smithsonian Institution does not know the relation of action to reaction, and of the need to have something better than a vacuum against which to react —to say that would be absurd. Of course he only seems to lack the knowledge ladled out daily in high schools. . . ."

Forty-nine years later the *Times* printed a retraction of this statement—on July 17, 1969, just before the United States landed a man on the moon.

But in all fairness, the *Times* had not been alone in criticizing Goddard. In fact, Goddard had received so much misunderstanding of and negative press reaction to his theories that he kept his work quiet for years for fear of further damaging his reputation.

OTHER ERRORS OF THE *TIMES*

The *New York Times* appears to have had a problem accommodating scientific progress. Not only did it react negatively to the possibility of space travel, but it also had difficulty with advances in the telephone and in flying.

Alexander Graham Bell demonstrated his telephone on March 10, 1876, and twelve days later the *New York Times* ran an editorial entitled "The Telephone." The only problem was that the editorial made no mention of Bell but did mention others, especially a German inventor Philipp Reis, for their work with the transmission of sound.

And then on December 10, 1903, the *Times* wrote about the futility of people's trying to fly and advised Samuel Langley, one of those working on a flying machine, to desist. Said the *Times:* "We hope that Professor Langley will not put his substantial greatness as a scientist in further peril by continuing to waste his time, and the money involved, in further airship experiments. Life is short, and he is capable of services to humanity incomparably greater than can be expected to result from trying to fly. . . . For students and investigators of the Langley type there are more useful employments."

December 10, 1903, the date of the newspaper in which this appeared, was one week before the Wright brothers made their first successful flight at Kitty Hawk.

FATAL NEWSPAPER ERRORS

Newspaper editors and reporters, in their rush to produce a daily publication and meet numerous deadlines, are vulnerable to many errors. One particularly embarrassing form of journalistic mistake is the printing of obituaries of people who have not really died. Three famous examples:

*Mark Twain's death was once widely reported while he was still very much alive. After he had heard about the publication of his obituary, Twain, in a cable sent from London to a New York newspaper on June 2, 1897, declared: "The reports of my death are greatly exaggerated."

*In 1960, Ernest Hemingway survived a plane crash in Africa—only to discover that newspapers around the world were reporting that he had died in the crash.

*Robert Graves, the British poet, was seriously wounded in action during World War I, but he was still very much alive when the *London Times* officially listed him as dead.

The Major Reference Works

Consider the major reference works in a library: the encyclopedia, the atlas, and the dictionary. Each in its own way is associated with error.

THE *ENCYCLOPAEDIA BRITANNICA* RIDDLED WITH ERROR? MORE THAN 600 ARTICLES IN ONE EDITION OUTDATED? SAY IT AIN'T SO, HARVEY EINBINDER

The *Encyclopaedia Britannica,* considered the leading general reference work in the English language and touted by its publishers as for two centuries "the reference standard of the world," is fraught with error. In fact, a physicist, Dr. Harvey Einbinder, spent five years in the late 1950s and early 1960s researching the reliability of the *Britannica* and found it had "so many defects" he decided to write a book about his findings.

Taking the 1958 and 1963 editions of the *Britannica,* Dr. Einbinder discovered myths presented as fact, legends presented as truth, errors in dates and statistics. He also found obsolete articles that were simply reprinted from editions of the *Britannica* as old as the ninth edition, which was issued 1875–1889.

He published the results of his research in a 390-page book entitled *The Myth of the Britannica* (Grove Press, 1964).

Dr. Einbinder started by examining science articles in the 1958 edition. "I discovered obvious flaws in the entries on Heat, Vaporization and the Compton Effect," he wrote. Dr. Einbinder then asked himself, "If there were errors in an exact science such as physics, what was likely to be the case in other fields where truth and fiction, opinion and evidence, could not be so readily separated? To answer this question, I inspected a large number of articles in different fields."

Among his findings:

The *Britannica* offered as true or without qualification long-discredited legends about John Smith's being saved by Pocahontas, the Black Hole of Calcutta, the exploits of Paul Bunyan, the ride of Paul Revere, Robinson Crusoe's island, the existence of William Tell, even the tale of George Washington and the cherry tree.

In statistics the 1958 edition, years after World War II, stated that the population of Warsaw was nearly 30 percent Jewish. In dates, several articles within one edition offered conflicting times on when Abraham lived. In zoology, the encyclopedia offered erroneous information about lemmings (they are not suicidal), wolves (they do not travel in packs), and Alaska blackfish (the oft-repeated statement that they can be frozen solid for the winter and still live is presented by the *Britannica* as fact, but this has never been substantiated by any experiments).

Included in Dr. Einbinder's book is a list of 666 articles of a half page or longer in the 1963 edition of the *Britannica* that can be traced back at least fifty years to previous editions, with no updating or corrections having been made. The list, wrote Dr. Einbinder, suggests "the extent of obsolete material" in what was then the latest printing of the *Encyclopaedia Britannica.*

One author who relied on the *Encyclopaedia Britannica* for information for his book only to learn too late that the information was erroneous was Alec Waugh, the British novelist and essayist. In a book of essays on the West Indies, *Love in the Caribbean* (Farrar, Straus and

Cudahy, 1958), Waugh writes: "If one cannot trust the *Encyclopaedia Britannica,* where can faith begin?"

Waugh had always wanted to visit Saba because of a reference in "that august authority" to the Dutch West Indian island, where live "the finest boatmakers in the Caribbean." He wanted to see how, since there is no beach, the boats were lowered over the side of a cliff, as related in the *Britannica.* But since Saba was hard to reach, Waugh had to rely on the entry in the *Britannica* when he came to write an earlier book about the West Indies.

To Waugh's surprise, following the publication of his book, he was informed by a reader that there was "no truth whatsoever" to what he wrote about Saba, and Waugh was referred to a *National Geographic* story in November 1940, by Charles W. Herbert. The article told how Saba had no natural timber and was not a site for boatmaking: Islanders could hardly be expected to lift massive timbers 1,500 feet up to the top of the cliff and then lower the completed boats down to the water.

At first unbelieving, Waugh later visited the island. "I am convinced," he writes in *Love in the Caribbean,* "that Mr. Herbert was right and that the *Encyclopaedia* was wrong." Waugh interviewed a number of the oldest people of Saba and no one could remember a time when even their grandparents talked of boats being lowered over the cliff by ropes.

Concludes Waugh: "Having been all over the island and examined its remarkable geological conformation, I doubt whether boats have ever been built on Saba."

One of the most widespread errors made about the *Britannica* is the popular belief that it is published in Great Britain. Although it originated there, the *Britannica* was bought by Americans in 1899, and ever since the famed fourteenth edition was published in the twentieth century, American contributors and American influences have predominated in the creation of the *Britannica.* In fact, since 1943, the *Britannica* has been owned by the University of Chicago. And they bought it from Sears, Roebuck, who had bought it in 1928.

ALAS, THE ATLAS

Here are some of the mistakes made over the centuries that are preserved today in any atlas:

CANARY ISLANDS:
WHAT ABOUT NAMING IT FOR THE BIRD DOG?

The Canary Islands in the Atlantic Ocean off the coast of Africa are misnamed. The Romans called them *Canariae insulae,* which means "dog islands." They used this name because of the many canines they found there. But for centuries people have mistakenly believed the islands were named for the canary bird. In fact, the canary bird, also found on these islands, gets its name from the Latin *canaria,* "pertaining to the dog."

GREENLAND:
THE ONLY THING GREEN WAS
THE MONEY ENVISIONED

Greenland, the largest island in the world, got its name not because it has the lush green fields the name implies but because people were purposely misled to believe that it had greenery. Viking explorers called the frozen island in the Arctic Circle Greenland to attract settlers, who otherwise might have been scared off. However, only the coastal areas turn green, and even this greening occurs solely during Greenland's brief summer. But long after the Vikings had passed and their subterfuge had been discovered, the erroneous label remained as the island's official name.

WEST INDIES:
AN ERROR OF MISDIRECTION

The islands that today constitute the West Indies got their name from Christopher Columbus, who mistakenly called the area Indies because he thought that they were part of the Indies islands of Asia.

CLEVELAND:
"THE MISTAKE BY THE LAKE"

Cleveland, the largest city in Ohio, one of the leading industrial centers of the United States, was founded in 1796 by a surveyor for the Connecticut Land Company. His name was Moses Cleaveland, and a village formed on the site he had surveyed along the southern shore of Lake Erie was named after him. But in 1831 a newspaper printer misspelled the name, dropping the *a* from *Cleaveland.* The village, now grown into a city, has been known as Cleveland—without the *a* —ever since.

Interestingly, residents of Cleveland lovingly refer to their city as "the mistake by the lake," but not necessarily because of the typographical error.

THE MOST COMMON ERROR ABOUT DICTIONARIES

As for the dictionary, many of them have one word too many. The word? *Webster.*

Many people think a dictionary with *Webster's* in its title is linked with the famed dictionary created by Noah Webster or is put out by the company that published Webster's work and is authorized to publish any further editions. Actually, since a book title cannot be copyrighted, anybody can put out a dictionary entitled *Webster's.* Many do. And many readers are misled.

Ironically Noah Webster's original dictionary did not have his name in the title. Rather, it was called *An American Dictionary of the English Language.*

Another error associated with Noah Webster's dictionary is the initial misjudgment of publishers. Webster had great difficulty finding anyone to back his dictionary. In 1825 while in London, he tried for weeks to locate a publisher, writing of his experience that "some think well of it . . . but how to get it published I do not know."

While our standard reference works contain error, a tour around the library reveals error in or about the Bible, literature, and other creative works.

THE FIVE MOST COMMON ERRORS MADE ABOUT THE BIBLE

The Bible is the most widely published, bought, and read book. Yet we continue to err in our perception of what the Good Book says. Here are some not-so-divine mistakes:

1. *Adam and Eve ate an apple in the Garden of Eden.*
   The exact fruit Adam and Eve ate from the Tree of
   Knowledge is never specified by the Bible. In fact,
   most biblical scholars agree it was *not* an apple.
   Among the fruits offered as possibilities: the grape,

apricot, and pear. One fruit considered most likely is the fig since the Bible states that Adam and Eve used the leaves of a fig tree to clothe themselves.

2. *Noah was told to take animals two by two into the Ark.* Actually in the original Hebrew Noah was told to take clean animals into the Ark "seven and seven" (Genesis 7:2–3). Only the unclean animals—those unfit for sacrifices after the Flood—were to be taken "two and two." A reading of the biblical passages shows that God's emphasis was on those animals that would be taken into the Ark "seven and seven."

3. *The Children of Israel passed through the Red Sea.* The body of water that parted for the Israelites fleeing Pharaoh is not the Red Sea. In the original Hebrew it is called the Reed Sea. Only a mistranslation turned it into the Red Sea. This is why scholars believe that the actual water traversed by the Israelites was not today's Red Sea, but north of it, toward the Mediterranean, where reeds dot the waters.

4. *Jonah was swallowed by a whale.* The Bible never states that a whale swallowed Jonah, only that "a large fish" gulped him down. In fact, this reference to a fish makes it impossible for a whale to have swallowed Jonah. A whale is not a fish. It is a mammal.

5. *The laws that Moses brought down from Mount Sinai were the Ten Commandments.* Nowhere in the Bible in its Hebraic text is there a mention of the Ten Commandments. This group of laws and precepts is actually called in the original Hebrew "ten words" in two different passages in the Bible (Deuteronomy 4:13 and 10:4).

Even works of art based on the Bible can't seem to get things straight. Consider a great piece of sculpture by Michelangelo and a great painting by Leonardo da Vinci.

MICHELANGELO'S *MOSES THE LAWGIVER*

The first time the Bible was translated from Hebrew into Greek a mistake was made in the description of Moses coming down from

Mount Sinai with the tablets of the Ten Commandments. In the Hebrew, Moses is said to have *karan,* a "ray of light" shining from his forehead. In the Greek translation by Aquila Ponticus in the second century, this read *keren,* or "horns," which were coming from the head of Moses. When Michelangelo was sculpting Moses for his monumental statue *Moses the Lawgiver,* the sculptor followed the Greek rendering of this passage and gave the leader of the Jews two horns on the top of his head. The mistranslated Bible, combined with Michelangelo's sculpture, helped feed centuries of Christian belief that Jews had horns, a belief that has persisted in some quarters even down to modern times.

DA VINCI'S *THE LAST SUPPER*

The Last Supper by Leonardo da Vinci, a scene which actually depicts a Passover seder, shows everyone sitting straight at a table. As a historian pointed out in a letter to *Time* magazine in 1981, this is historically inaccurate. Jews at a seder sit in a reclining position on their left side throughout both the service and the meal to show their freedom from slavery. Another painter in his rendering of the Last Supper showed rolls on the table (instead of the required unleavened bread, matzoh; bread is forbidden during Passover). And a French artist once depicted his *Last Supper* having a table set with cigar lighters.

NOTABLE SLIPS BY PUBLISHERS AND CRITICS

The history of literary work is replete with instances of critics, publishers, and others making costly, embarrassing mistakes in assessing the merit of creative products. Some of the most outstanding examples of slips made with rejection slips:

In Book Publishing

James Joyce: The famed Irish author saw his book of short stories *Dubliners,* now considered a classic, rejected by twenty-two publishers.

Hans Christian Andersen: The first volume of his *Fairy Tales* was turned down by every publisher in Copenhagen. He had to publish it at his own expense.

Harriet Beecher Stowe: Her *Uncle Tom's Cabin,* originally a serial in an antislavery publication, was offered to a publisher whose reader and critic advised against publishing it in book form because it lacked general interest. The wife of the critic, though, liked it and pushed for publication.

Daniel Defoe: His *Robinson Crusoe* was offered to publisher after publisher but returned each time.

William Makepeace Thackeray: He wrote *Vanity Fair* for a magazine, but a number of publishers declined to publish it in book form, saying it had little interest. Thackeray finally had to publish it himself in monthly installments.

John Creasey: The author, under numerous pseudonyms, of 564 published books was also the recipient of 743 rejection slips before he had one of his mystery novels accepted.

Marianne Moore: The American poetess had one of her poems rejected thirty-five times.

Pearl Buck: The Nobel Prizewinner had her novel *The Good Earth* turned down twelve times.

J. P. Donleavy: His *The Ginger Man* was rejected thirty-six times.

Frank O'Connor: This short story writer once submitted one of his works to a magazine, which accepted and paid for it; later, a second editor, finding the manuscript on his desk, read the story and sent O'Connor a rejection letter.

In Plays

Fiddler on the Roof: The first review of *Fiddler on the Roof,* which was to become at one time the longest-running musical in Broadway history, was a negative one by the esteemed *New York Herald Tribune* drama critic Walter Kerr, now drama critic for the *New York Times.* In his review, which appeared in the first newspaper to be brought into the theater party then awaiting the reviews, he wrote that he thought *Fiddler* "might be an altogether charming musical if only the people of Anatevka did not pause every now and again to give their regards to Broadway, with remembrances to Herald Square." Kerr went on about how the play dipped from its potential "by touching character too casually," and he referred to the play's "easy quips," "lyrics that stray too far from the land," and its "occasional high-pressure outbursts that are merely marketable." He concluded: "I very much miss what it might have been."

In a retelling of this scene in the book *The Making of a Musical* (Crown, 1971), Richard Altman, assistant to the play's director, wrote how Kerr's review was "devastating" and how it "wrapped a cloak of melancholy" around the room. Jerry Bock, who wrote the musical, and Sheldon Harnick, who wrote the lyrics, "left immediately, and before long other people began drifting away, myself among them. The joy was gone; the bubbles had left the champagne; it seemed best just to go home." It was not until the next morning that other reviews praising the show came out, and with lines forming at the box office—by midmorning it took two and a half hours of waiting to buy a ticket—*Fiddler*'s success was finally assured.

Life with Father: Producer Oscar Serlin found no one who would share his enthusiasm for a play based on the *New Yorker* magazine stories of Clarence Day, Jr., about growing up with his father, a stockbroker, in the New York of the turn of the century. Desperate, Serlin sneaked a copy of the play into the suitcase of wealthy John Hay Whitney, an investor in theater productions. But Whitney's adviser on the theater, Robert Benchley, was against it, stating, "I could smell it as the postman came whistling down the lane. Don't put a dime in it." (Fortunately Whitney did not make the mistake his adviser did and decided to back the play anyway. It went on to run for seven and a half years and 3,224 performances on Broadway—the longest-running play until that time.)

But play backers were not the only ones who initially vetoed *Life with Father*. Actor Walter Huston twice turned down the role of Father, and the famed acting team of Alfred Lunt and Lynn Fontanne also refused any part of the production.

Never Too Late: which ran on Broadway for 1,007 performances, was rejected by at least twenty-five New York managements (including David Merrick). But only after another six years of making the rounds, and with several out of New York State productions (including one with an unknown named Dick Van Dyke), did this comedy about a middle-aged couple, with grown children, learning they are about to be parents again find its way onto Broadway. The final director? The noted George Abbott—whose office had been one of the original twenty-five naysayers.

Man of La Mancha: Richard Barr, producer of *Who's Afraid of Virginia Woolf?* and *Sweeney Todd,* turned down the offer to become involved in producing *Man of La Mancha.* Later, offered a chance to produce a musical called *Drat,* he decided not to pass up this chance. *Drat* closed in one night.

Come Blow Your Horn: Neil Simon's first comedy was turned down by Max Gordon, one of the theater's great producers *(Born Yesterday, The Solid Gold Cadillac, The Late George Apley).* He always lamented what this mistake meant: He missed out on producing not only *Come Blow Your Horn* but the rest of Simon's unbelievable string of hits.

West Side Story: Cheryl Crawford, the producer, was offered the chance in 1956 to produce a play being worked on by Leonard Bernstein, Stephen Sondheim, Arthur Laurents, and Jerome Robbins. The work involved converting *Romeo and Juliet* into a musical. Twenty prospective backers, however, turned thumbs down on the play at a special audition for them. Richard Rodgers and Oscar Hammerstein II, asked to comment on the musical, observed that they doubted enough youthful actors who could sing the score could be found. Crawford finally backed out, too, deciding that the production was too expensive and would not work. *West Side Story,* though, eventually reached Broadway—and much success. The leading drama critic of the day, however, yawned at the outcome, saying he had found the musical "almost never emotionally affecting." His name? Walter Kerr.

HOW THE WORLD'S MOST POPULAR SONG SPENT THREE YEARS IN A TRUNK

Sometimes the creative artist himself errs in evaluating his creation. This happened with nothing less than the most commercially successful song ever written.

In 1939 Irving Berlin composed "White Christmas" but thought so little of it that he tossed it into a trunk. He did not see fit to retrieve it from there until he needed it for a Bing Crosby-Fred Astaire movie, *Holiday Inn.* Crosby, a staunch Catholic, at first resisted singing the song because he felt it tended to commercialize Christmas. He finally agreed, took eighteen minutes to make the recording, and soon saw the song, especially his version of it, become an all-time hit. Crosby's version has sold more than 25 million copies. All together, "White Christmas" has appeared in 550 versions, selling 6 million copies of sheet music and 136,260,000 records—just in the United States and Canada.

Berlin, however, was saved by a friend from making a glaring mistake with "White Christmas." His opening line for the song was originally "I'm sitting by a pool in Beverly Hills dreaming of a White

Christmas." The friend prevailed on him to drop the Beverly Hills reference as "inappropriate."

"IF AT FIRST . . ."

Here are more literary greats who encountered less than great critical receptions during their lifetimes:

*Tennessee Williams, the American playwright, saw his first play, *Battle of Angels,* so poorly received that it was roundly booed. The producers came forward and apologized to the audience.

*Charles Lamb, the British essayist and critic, had his first play hissed off the stage. Seated in the audience, he actually hissed along with the audience so that he would not be recognized as the author.

*Rudyard Kipling, who won the Nobel Prize in literature, was fired from his first job as a reporter for the *San Francisco Examiner.* "This isn't a kindergarten for amateur writers," his editor told him. "Sorry, Mr. Kipling, but you just don't know how to use the English language."

*Once an apprentice to an apothecary, the poet John Keats was advised by one critic to go "back to plasters, pills and ointment boxes"—with the caution to "be a little more sparing of extenuatives and soporifics in your practice than you have been in your poetry." Keats was so poorly received as a poet during his lifetime that he asked that on his gravestone be engraved, "Here Lies One Whose Name Was Writ in Water." Keats, who died at the age of twenty-five, was buried with the epitaph on his tombstone.

*Herman Melville's *Moby Dick* was originally reviewed by the *New Monthly Magazine* as "maniacal . . . gibberish, screaming, like an incurable Bedlamite, reckless of keeper or strait-waistcoat."

*William Wordsworth, the English poet, saw his poems termed by critics "silly," "infantile," "nauseating," "disgusting."

*Walt Whitman's *Leaves of Grass* was assaulted by

217

critics, who hurled such invectives at it as "entirely
bestial," "slopbucket," "rotten garbage," and
"venomously malignant."

WHEN A PUBLISHER REJECTED A NOVEL IT HAD ALREADY PUBLISHED

Even an award-winning book once successfully published was later
rejected.

In 1977 Chuck Ross, a struggling, frustrated writer, decided to try
an experiment. He typed up a fresh manuscript copy of Jerzy Kosinski's
acclaimed novel *Steps,* changed the title, and submitted the work under
his by-line to fourteen publishers. All fourteen rejected the novel that
had won the National Book Award in 1969 for best work of fiction.
Among the publishers turning down the manuscript was Random
House—the book's original publisher.

Mistakes have affected not only the work but the lives of famous
literary lights. Here are three examples:

THE DAY COLERIDGE SHOULD HAVE STAYED IN BED

Samuel Taylor Coleridge, the English poet, lost two-thirds of his most
memorable poem because of a mistake. In 1797, while convalescing
after an illness, he began reading about Kubla Khan's palace and fell
asleep from the drugs he had been taking. When he awoke, he realized
that he had written in his sleep a long poem about Kubla Khan. He
envisioned a 300-line work and immediately began writing it down. He
got through 54 lines when there was a knock at the door. Instead of
finishing the poem or asking the visitor to wait, Coleridge invited his
visitor in and spent more than an hour talking to him. When the visitor
finally left and Coleridge returned to his writing, he realized he had
forgotten the rest of the poem. He never was able to finish what came
to be called "Kubla Khan," one of the best-known poems of English
literature but an unfinished work of only 54 lines.

WHEN A PRINTER NAMED A NOBEL PRIZEWINNER

A printer's typo changed the name of a future Nobel Prizewinner.
William Faulkner, the American novelist, was born William Falkner—
of a long line of Falkners, all without a *u* in their last name. But when

his first book was published, he found that the printer had added a *u* to make it Faulkner. He decided it was easier to switch than fight, and thereafter the author of some of the most significant fictional works in American literature signed his name and his books as Faulkner.

AND THE WINNER IS . . .

Authors who were never awarded the Nobel Prize in literature

Henrik Ibsen
Henry James
Émile Zola
Joseph Conrad
Leo Tolstoy

Authors who were

Giosuè Carducci
Henryk Sienkiewicz
Rudolf Eucken
Selma Ottiliana Lovisa Lagerlöf

THE MISTAKE THAT HELPED A WRITER

Not all writers have been the victims of mistakes. One author used a mistake to his lifelong benefit. Voltaire, the French writer and philosopher, was helped in his career by a mistake in arithmetic. A brilliant mathematician, Voltaire noticed that a lottery being run by the French government contained a miscalculation which could enable someone who understood the error to capitalize on it in playing the lottery. He formed a syndicate, bought up all the tickets, and saw his concept prove true. On the basis of his winning share, he became wealthy enough to live independently the rest of his life—and to use this freedom to write prolifically on a wide range of topics.

One would think that at least our fairy tales would be immune from error. After all, how can out-and-out fantasy be incorrect? Yet probably the best-known fairy tale of all time is built around a 300-year-old mistake.

THE MISTRANSLATION THAT CHANGED CINDERELLA'S SLIPPER INTO GLASS

The earliest version of the Cinderella fairy tale says that her slippers were made out of gray and white squirrel (*vair* in the original). But this

219

word, which is used in heraldry, was obsolete by 1697, the time a Frenchman, Charles Perrault, was writing one of the earliest Mother Goose collections of fairy tales. In his version of Cinderella, Perrault used the French *verre,* meaning "glass." Thereafter Perrault's interpretation became the French version of the story in which reference was always made to Cinderella's glass slippers.

But this version and those based on it are the only ones to talk of glass slippers. In the Cinderella story told in other languages, the slippers are referred to as being made out of silver, satin, pearls, or jewels. The version in *Grimm's Fairy Tales,* which originally appeared in German and is among the most famous of the Cinderella stories, never mentions glass slippers at all. According to *Grimm*'s "Cinderella," which is reprinted in the Harvard Classics as an example of a classic fairy tale, the slippers are made of gold.

THE CHINA SYNDROME

Should we have a more understanding attitude toward error in our libraries? Actually the Chinese do. The Chinese scholar and writer Lin Yutang (1895–1976), in an essay on the differences between Americans and Chinese, touched on the Chinese approach to errors in books. Observed Yutang: "An American editor worries his hair gray to see that no typographical mistakes appear on his pages. The Chinese editor is wiser than that—he leaves his readers the supreme satisfaction of discovering a few typographical mistakes for themselves."

"TRUTH IS OFTEN SAID IN ERROR" DEPARTMENT:

TYPOS TO LEARN BY

While "a truth is often said in jest," many times a truth is also inadvertently revealed by error. Here are just some examples of typos that reveal, often by a mistake with just one letter, more truth than originally intended:

— The *Reading* (Pennsylvania) *Eagle* reported, "The local Conrail office has announced a half-fair policy for senior citizens."

— A mortician was quoted in the *Journal,* published in Flint, Michigan, as saying, "I have never urned any family away because they couldn't pay for a funeral."

— A classified ad in the *News-Citizen,* published in Vandergrift, Pennsylvania, stated: "FREE TO GOOD HOME—part wire-haired terror."

— The *Dayton* (Ohio) *News* ran an ad for grass seed with this description: "Produces a lawn for all-purpose use. Dies well in full sun or partial shade."

— The *Albuquerque* (New Mexico) *Journal,* intending to provide information about patents, instead related that "a parent in the United States is good for 17 years." Which is just about right.

— An Associated Press dispatch out of New York City told of the soldier who "has been in the Army 12 years and has risen to the rant of sergeant."

— The *St. Louis Post-Dispatch* revealed that a congressional subcommittee was "investigating charges that servicemen are used to perform mental tasks for officers."

— The *New York Journal-American,* which has long ceased publication, may have contributed to its own demise with such tidbits as the following business news: "There was something about that title, Old Incestors' Trading Corporation, that inspired confidence."

—A high school newspaper in Canton, Ohio—the *Eagle's View*—revealed to its readership that "Mr. Harris has sad 12 years teaching experience."

—Small-town newspapers provide their communities with close personal coverage, such as the *Pittsburg* (Kansas) *Headlight* social item that "following the ceremony, a wedding supper was hell at the home." And the Lancaster, Pennsylvania, paper *New Era* quoted the parent who lamented, "Valeria is 17 and giving us nothing but hearache."

CHAPTER VIII
Faux Pas: Error in Daily Life

*Don't rush me. I'm making misteaks as fast
as I can.*

—Sign over a secretary's desk

In modern times, errors and accidents have become an ever-more-frustrating part of the world in which we live, work, and play. Some of the increase in goof-ups and breakdowns can be seen as a sign of deterioration in workmanship, materials, and service. Other mistakes are evidence that our machines and computers are really no better than the human beings who run them and program them. Garbage in, garbage out.

During 1982 advice columnist Ann Landers published a letter from one of her readers which enumerated some of the foul-ups the correspondent had encountered in daily life. The litany is instructive of what many people are now enduring just to get through the day: "Last month I caught two errors on my bank statement, three mistakes on my insurance forms (one made by the insurance company, two by the doctor's office), an error on my water bill, two errors on a listing that involved my car (made by the Department of Transportation). . . . I receive an airline ticket with a wrong date, and the flight number is goofed up. . . . This morning a button came off a new suit the first time I wore it and the door on my 1982 car wouldn't open from the inside."

The writer, who signed the letter "Had It in the Nation's Capital," asked if this indicated that people didn't care about quality anymore.

Ann Landers told "Had It" that actually these complaints were some of the less serious ones she had heard, that "the mistakes physicians and dentists make are hair-raising, not to mention lawyers, auto mechanics, builders, plumbers and electricians."

The columnist ascribed the increase in errors to a decrease in the craftsmanship in these times and the resultant lack of people who take pride in their work.

"Can it be," Ann Landers asked, "that too many things are being done by machines and people don't care anymore?" And then she answered her own question. "I hope not, but that's the way it's beginning to look."

Indeed, the modern world is relying more and more on machines —especially the ultimate machine, the computer. The results many times are not only more productivity but also mistakes more difficult to correct and not only lower costs for companies but higher frustration for customers.

To counter this problem, scientists have begun developing a new field: human factors engineering. First emerging as a separate scientific discipline during World War II, human factors engineering utilizes the multidiscipline field of engineering and psychology so that the capabilities and limitations of people are factored into the design, construction, and operation of equipment, machines, and buildings. The purpose is to achieve maximum performance by reducing human error.

The first human factors problem to be recognized occurred during World War II, when it was realized—only after frequent landing accidents—that pilots trained on the B-25 bomber mistakenly used the wrong controls when trying to land the B-26. The reason: The landing gear and wing-flap levers were positioned differently on the two planes.

A similar problem also emerged in the flying of fighter planes. The levers for the adjustment of the seat and for the ejection seat were positioned in such a way that many pilots, reaching for the seat adjustment lever, often accidentally activated the ejection seat. The result: the ejection of a very surprised pilot; the destruction of a very expensive airplane.

One aspect of our daily life in which human factors are receiving careful attention is today's nuclear power plants. The 1979 accident at the Three Mile Island nuclear reactor has made the reassessment of the design and operation of this and other nuclear facilities vitally necessary.

For example, the Special Inquiry Group of the Nuclear Regulatory Commission discovered that although the Three Mile Island control room had been designed with 750 alarms to signal trouble, these alarms proved of little use the day of the accident because so many of them were flashing and because their placement in the control room was random. Ironically, designers had placed the very instruments relevant for assessing this particular accident on the back of the panels of the consoles—where the operators could not readily see them.

And another important gauge was later found to have a tag sitting over it, obscuring its reading.

Dr. Robert W. Deutsch is president of General Physics, a Washington, D.C., area company which added a human factors engineering division in 1979. In a booklet on nuclear power, he wrote: "The Three Mile Island experience proved that no matter how carefully a power plant is designed to operate in theory, it is impossible to totally eliminate human error."

Mutual of New York [one of America's largest insurance companies] estimates that 70 percent of its dictated correspondence has to be redone at least once because of errors.

—William McGowan, "Iliterasee att Wurk,"
New York Times Op-Ed page,
August 19, 1982

Pull the Plug: A Look at Computer Error

The computer is shaping the modern world. But with computers has come computer error. Except that computer error is really human error. Computers are no better or worse than the data that are fed into and extracted from them by people. As we have seen, whenever humans get involved, error evolves. The product of computers is becoming bent, spindled, and mutilated because of the spreading terror of error.

SORRY ABOUT THAT NUCLEAR ATTACK, SIR

Computers are involved with the United States' nuclear arsenals, providing rapid input and decision making. Thus, if these computers make a mistake, the potential for damage is enormous. Yet during one eighteen-month period the North American Air Defense Command had 151 false alarms because of computer error. Four resulted in orders that increased the state of alert of B-52 bomber crews and intercontinental ballistic missile units. A major false alert that lasted six minutes happened when a technician, by mistake, placed on an American military computer a training tape of a Soviet attack.

THAT'S WHY IT'S NOW CALLED SOCIAL INSECURITY

An important study of a vital government program was seriously affected by a computer foul-up.

In 1974 Martin S. Feldstein, a Harvard professor then head of the National Bureau of Economic Research in Cambridge and at present chairman of the President's Council of Economic Advisors, completed a pioneering study of Social Security and how it affected or did not affect private savings. He concluded that Social Security reduced personal saving by 50 percent, with serious consequences for capital formation and output.

Six years after that study had been released, two economists in the Social Security Administration, Dean R. Leimer and Selig D. Lesnoy, announced that they had discovered a fundamental flaw in Feldstein's study. This in turn had led to the discovery that Feldstein had used a computer program which erred in repeating a calculation over and over. Consequently, the data Feldstein drew on showed an overestimation of the value consumers place on Social Security.

The finding of this computer foul-up "stirred a ruckus in the world of economics," wrote the *New York Times* in an October 5, 1980, story entitled "Martin Feldstein's Computer Error." As a result, the *Times* reported, the Feldstein conclusion about Social Security's impact on private sector savings "cannot be proven or disproven on the basis of available work."

In his defense, Feldstein pointed out that his critics themselves erred in leaving out of their model some crucial 1972 legislation that raised the benefits by 20 percent and indexed these higher benefits against inflation. Feldstein, however, admitted he, too, had left this legislation out of his calculations. Once it was now figured in, he said, his original findings were justified, although still not in the proportion originally expressed.

But the computer program which repeated a calculation over and over had left Feldstein's work up in the air and subject to doubt.*

The effects of a non-performing computer can go beyond frustrated expectations. A small business can become so critically dependent on a computer for its billing and accounting that, if the computer errs, the business can go bankrupt without even realizing it.

—Andrew Pollack,
"When Computers Don't Work,"
New York Times, October 23, 1981

AND NOW THE WINNER ISN'T . . .

Sometimes the problem is not the programming of the computer but the reading of what the computer puts out.

In 1981 the election contest between the parties of Israeli Prime Minister Menachem Begin and former Defense Minister Shimon Peres was a closely fought campaign eagerly watched not only in Israel but around the world. Hanoch Smith, an American-born Israeli pollster

---

*The *New York Times* asked another economist if "mistakes of such magnitude" are often made by economists. Responded Alan Blinder, a professor at Princeton who was doing his own studies of Social Security, "There are probably untold numbers of errors in the economic literature."

who had correctly predicted the previous national election in 1977, was now being asked by major world television networks and the international press to predict the outcome.

When the polls closed, Smith revealed on television the results of his election day sample—a sample that in 1977 had been astonishingly accurate.

But then, "horror of horrors," as one news service termed it, Smith got it wrong not once but twice. Before midnight and again after midnight, he announced that computer projections were showing that Menachem Begin's Likud party had lost seats and that Shimon Peres's Labor alignment would form the next Israeli government.

On the basis of these findings, Peres went before the international news media and proclaimed victory. The news of his acceptance speech and Begin's defeat was reported around the world.

The only problem was that when all the ballots were counted, the opposite was true, and Begin, the feisty prime minister, had been retained in office.

When Smith's error became known, both Peres and Begin, as well as their supporters, expressed anger at the foul-up and confusion.

Later interviewed to find out what had happened, Smith was, according to news accounts, relaxed and unperturbed. "For me it's a bit of fun to be in the limelight," he said. To explain the mess-up, he pointed to human error in handling the computers. "I'm angry about the mistakes which arose simply because the computer programmers misread the data.

"Next time," he vowed, "Israel TV has promised me that I can select my own computer services and everyone can rest assured that there'll be no blunders."

Smith, a native of Pittsburgh, had been correctly forecasting elections since 1957—until the summer of 1981, when he and a computer service sent former Israel Defense Minister Shimon Peres out before the world to proclaim a victory that did not exist.

WHEN BANK COMPUTERS ERR, HEAVEN HELP US

Stanton Powers had only $1.17 in his bank account, but he had a lot of faith in his heart. On the advice of a friend he decided to pray to God for help by using a special meditation. He also decided to pray in a special place—before an automated teller machine at the County Bank of Santa Cruz in California.

On September 7, 1982, the thirty-nine-year-old artist living on Social Security disability payments began praying hard and long in front of the machine.

As Powers later told bank and police officials, he soon began to see his bank balance on the computer screen change. When he finished praying that night, the account had grown to $1,600.

He went home and tried to sleep, but it seems he could find neither solace nor rest. At 5:30 A.M. he went back to the bank and began praying again before the machine. When he finished, his account stood at $4,443,642.71. Powers quickly withdrew $2,000.

As soon as the bank learned of Powers's power of prayer, they immediately froze the multimillion-dollar account and began an investigation. Powers went and got himself a lawyer, who announced, "Not being prepared to deny the existence of God, I am requesting that the bank show us empirical evidence that it was an error."

The bank's initial study of its records indicated that a computer error—not a miracle—was the cause of the sudden increase in the Powers account. Since no money could be found to have been deposited into the account, the numbers represented just that—numbers, not actual money.

How then did Powers, who denied punching falsified numbers into the system, get $2,000 out of his nearly bare account?

The bank's lawyer did not have the answer to that. "I'm not prepared to discount acts of God in any way," he said. "But I really doubt that's a factor in this matter."

The lawyer for Powers said his client was holding on to the $2,000 while the bank continued its search of records. He expressed doubt that his client could be charged with a crime for keeping the money.

"I don't see how Mr. Powers could be charged with anything other than being a very religious man," his lawyer said. "The law would have to decree that miracles can't happen. I don't think the American justice system is prepared to extend into the realm of the deity."

Deity or not, it was obvious that because of some foul-up or gremlin somewhere, a computer had at long last achieved what many had feared: It had become a new god that at least one person now worshiped.

DEAR MR. MURDERER

During the first year of Ronald Reagan's presidency the Republican National Committee sent out a computer-generated letter over the signature of President Reagan asking for generous donations for the Republican cause. One recipient on the mailing list was Mr. Wilbur May. The address was somewhat odd, however: H-1-8, Jackson, Georgia.

This turned out to be a cell in the Jackson prison. It also happened to be not just an ordinary cell but a cell in the death house. The occupant of the cell, whose name was really Brandon Astor Jones, was awaiting execution when he received the warm missive from the president, who invited him to participate more closely in the American democratic process. "I only wish I could call you as a witness" to the great national debate over the future, wrote Mr. Reagan.

Since Jones had a prior engagement, he sent the letter on to a friend —after having had, according to the friend of the condemned prisoner, his first good laugh in a while.

We must remember that computers process error at the same bewildering speed at which they process truth.

—A General Electric computer executive

Error also invades our daily lives in an insidious, far-reaching and often fatal way—through accidents.

Although not all accidents are caused by human error (weather, structural defects, and unforeseen situations are also causes), the human element contributes to the vast majority of accidents. Even alcohol, said to be a factor in half of traffic fatalities, causes auto crashes because it dulls human faculties, leading drivers to err in handling their cars (taking a drink before driving may therefore be the biggest driver error of all).

The deadly consequences of error in our lives can be seen in statistics compiled by the National Safety Council. Accidents are the leading cause of death among all persons from one to thirty-eight years of age (between the ages of fifteen to twenty-four, accidents claim more lives than all other causes combined). In the United States today only cancer, heart disease, and stroke kill more people overall than accidents.

The following is a look at how accidents have pervaded our lives. Unless otherwise stated, accident statistics are those recorded in the United States in 1980, the first year of this decade and, as of this writing, the most recent full year in which statistics were available. During those 366 days, 70 million people were injured, 10 million were disabled, and 105,000 died because of accidents. The largest cause of accidental fatalities? Motor vehicles, which killed 52,411 people. The cost of accidents to the nation was $83.2 billion. Error does not come cheap.

Famous Firsts

THE UNITED STATES' FIRST AUTO FATALITY WAS BLISS

On September 14, 1899, in New York City Henry H. Bliss, a 68-year-old real estate broker, stepped off a trolley and into history. He descended from the streetcar, turned to assist a woman passenger, and was hit by—what else?—a cab. Bliss became the first person to be killed in an auto accident in the United States.

GUESS WHO WAS THE PILOT IN FIRST FATAL PLANE CRASH?

The first plane accident to claim a life occurred on September 17, 1908, five years after the first successful flight by Orville and Wilbur Wright. The pilot in that fatal plane crash was none other than Orville Wright. He was accompanied on the flight by Lieutenant Thomas E. Selfridge of the U.S. Signal Corps. In mid-flight, the propeller broke, the plane plunged 150 feet, and Selfridge was killed. Orville suffered multiple hip and leg fractures.

A FAULTY PARACHUTE CAUSED FIRST FATALITY IN SPACE

Russian cosmonaut Vladimir Komarov was the first person to die in space. He was launched into space on April 23, 1967, from Tyura Tam, USSR, aboard the *Soyuz 1* spacecraft. His mission was a twenty-five-hour eighteen-orbit flight, after which his capsule was to return to earth suspended by parachutes. Komarov died when the straps of the parachutes broke and the capsule plummeted to earth.

The first fatalities in the American space program occured January

27, 1967, when three astronauts were killed in a flash fire as they sat in their *Apollo* spacecraft during a routine ground test. An investigation revealed that "113 significant engineering orders" had not been correctly carried out in the construction of the *Apollo* capsule. As a result of these findings, the capsule was extensively redesigned.

FIRST ACCIDENT INSURANCE POLICY IN AMERICA

The idea that people could insure themselves against the pain, suffering, and loss of wages caused by error came in the 1860s. The Travelers Insurance Company issued the first accident policy in 1864 to James Bolter of Hartford, Connecticut. He bought a $1,000 policy to cover him on his walk from his job at the Post Office to his home on Buckingham Street. The cost of the premium: 2 cents.

Error on the Road

As of the beginning of 1981, approximately 2,340,000 people had died in traffic accidents in America since the United States began keeping records. It took almost fifty-two years for the first 1 million fatalities to occur, just twenty-two years for the 2 millionth. The National Safety Council estimates that the 3 millionth motor vehicle death will occur by the early 1990s, a less than twenty-year time span. It seems that death, like life, is becoming increasingly frenetic.

THE TOP TEN STATES WHERE YOU TAKE YOUR LIFE IN YOUR OWN HANDS ON THE ROAD

The most dangerous state in which to drive a car is intoxicated—and Wyoming. According to the National Safety Council, in 1980 Wyoming had 51.7 traffic fatalities per 100,000 population. Here are the top—or bottom—ten states for driver and pedestrian fatalities:

1. Wyoming	51.7*
2. New Mexico	47.2
3. Nevada	43.2
4. Montana	41.2
5. Arizona	34.7
6. Idaho	34.7

*Motor vehicle traffic deaths per 100,000 population.

7. South Dakota	32.9
8. Oklahoma	32.0
9. Texas	31.0
10. Florida	29.5

The state with the lowest rate is Rhode Island with 13.5. The U.S. rate is 23.2.

SOME SURPRISES ABOUT MOTOR VEHICLE STATISTICS

WHERE

With all the talk about city traffic congestion and accidents, the surprising truth is that almost two out of three deaths happen not in cities or even in urban areas but in places classified as rural.

WHO

In a city more than a third of the victims of a fatal motor vehicle accident are pedestrians. In rural areas, victims are almost always (by fifteen to one) car occupants.

WHEN

More than half of all fatal vehicle accidents—whether in the city or in the countryside—happen at night. In fact, the mileage death rates at night are nearly four times the day rates.

The most likely time to have a fatal accident: 1:00 A.M. on a Sunday (12.1 percent).

Least likely: 5:00 A.M. on a Friday (1.2 percent).

Most likely time for all accidents: 3:00 P.M. Friday (8.9 percent).

THE TEN MOST DANGEROUS TURNPIKES

In 1980 the mileage death rate on the nation's turnpikes was 1.2 per 100 million vehicle miles. But it was far safer to be on a turnpike than on a rural road, where the death rate was 5 per 100 million vehicle miles.

	Deaths per 100 million vehicle miles
1. New York State Thruway	46
2. New Jersey Turnpike	40
3. Florida Turnpike	36
4. Garden State Parkway (New Jersey)	28
5. Oklahoma Turnpike	28
6. Ohio Turnpike	26
7. Illinois Tollway	20
8. Pennsylvania Turnpike	18
9. Indiana East-West	17
10. Massachusetts Turnpike	12

STAY OFF THE TWO-LANE HIGHWAY

In 1980, 68 percent of all city motor vehicle accidents and 64 percent of all injuries occurred on two-lane roads.

JUST THOUGHT YOU'D LIKE TO KNOW . . .

Trees are fifteen times more likely to be involved in an accident than fire hydrants.

TRAFFIC TIE-UP

The human being and the automobile seem to be a combination of errors waiting to collide. In the entire state of Ohio in 1895 only two cars were on the road. Their drivers crashed into each other.

THE ACCIDENTAL ROAD SIGN

There is a town in Maryland called Accident. It sits on a major state highway linking western Maryland with the rest of the state. On the approach to the town is a road sign that doubles as a warning and is always true, no matter what the traffic condition. The sign says: ACCIDENT AHEAD.

CAN EDUCATION CUT DOWN ON DRIVER ERROR?

According to the National Highway Traffic Safety Administration in a study released in 1981, "no statistically significant difference" could be found in accident rates between those who had or had not taken driver education.

THE MOST DANGEROUS PLACE TO BE

"Home Sweet Home" should really be "Home Dangerous Home," for in terms of injuries the home is the most dangerous place to be. More than 4 million Americans are injured each year in home accidents. And 28,500 die from them. Half of all these deaths are caused by falls. Burns come next in frequency (more than 70 percent of fatalities from fires occurred in homes).

THE WEAKER SEX?

In every category of accident—motor vehicle, falls, drownings, fires, choking, firearms, swallowing poison, poison by gas—more males die than females. In fact, 70 percent of all accident victims are males.

THE MOST DANGEROUS MONTH

Beware the Ides of July. This is the most "accident-prone" month. In July 1980, 11,600 accidental deaths were recorded, far above the monthly average of 8,750. You were more likely to die in your house in December and February (which account for 21 percent of home accidental fatalities) but far more likely to die from motor vehicle accidents and drowning in July.

HOW PEOPLE DIED ACCIDENTALLY IN ONE YEAR

A study of accidental deaths in 1980 showed that of the total of 105,000 such fatalities, the categories and their death rates were:

		46.2
		(rate per 100,000
All accidents	*105,000*	*population)*
Motor vehicles	52,600	23.2
Falls	12,000	5.4
Drowning	7,000	3.1

		46.2
		(rate per 100,000
All accidents	*105,000*	*population)*

Fires, burns, etc.	5,500	2.4
Suffocation/choking	3,100	1.4
Poisons (by solids, liquids)	2,800	1.2
Firearms accidents	1,800	0.8
Poisoning (by gases and vapors)	1,500	0.7
All other types (falling objects, air travel, medical complications)	18,400	8.0

SOURCES OF MOST FREQUENT ACCIDENTAL INJURIES

Below is a list of the products or activities that caused the injuries reported most frequently by hospital emergency rooms, according to the Consumer Product Safety Commission's National Electronic Injury Surveillance System. Automobile accidents are not included. The column at right is the CPSC's extrapolation of the number of injuries nationwide from July 1, 1980, through June 30, 1981.

Stairs	763,000
Bicycles and accessories	518,000
Baseball	478,000
Football	470,000
Basketball	434,000
Nails, carpet tacks, screws	244,000
Chairs, sofas, and sofa beds	236,000
Skating	225,000
Nonglass tables	225,000
Glass doors, windows, and panels	208,000
Beds	199,000
Playground equipment	165,000

Lumber	151,000
Cutlery and knives (unpowered)	140,000
Glass bottles and jars	140,000
Desks, cabinets, bookcases	126,000
Swimming	126,000
Drinking glasses	111,000
Ladders and stools	99,000
Fences	99,000

THE UNFRIENDLY SKIES: AIRPLANE ACCIDENTS

General aviation—which includes all civilian aircraft except airliners—claims an average of 27 lives a week. The total number of lethal accidents has held at more than 675 each year. Weather and mechanical defects cause some accidents, but according to a *Wall Street Journal* article on safety in the skies, crew error is blamed for nine out of ten fatal air crashes.

In 1977, to test for pilot error, NASA simulated a flight between New York and London. Included in the flight simulation were several

---

HALF OF ALL PREGNANCIES ARE MISTAKES

While many deaths are caused by mistakes, so, too, it seems, are many births.

A 1983 study by the Alan Guttmacher Institute discovered that more than half of the 6 million pregnancies that occur each year in the United States are unintentional. And as evidence of how unintended those pregnancies are, statistics show that nearly half of the 3.3 million unwanted pregnancies are terminated by abortion.

One of the culprits in unwanted pregnancies appears to be a faulty use of the diaphragm as a contraceptive method. The Guttmacher study found that 18.6 percent of women relying on the diaphragm in the first year of marriage became pregnant. The report concluded that such contraception could be more successful, but only if "properly used."

---

emergencies and a few unusual maneuvers. At the end of the test NASA found that the pilots committed mistakes at the rate of 11.4 per hour.

A previous study, undertaken by a European airline in 1971, used an actual flight between London and Glasgow and back. Analyzing pilots under a high-workload situation (in such a short flight, most of the time is spent taking off and landing), the airline found that the rate of pilot error averaged 15.6 an hour.

THE TEN MOST DANGEROUS CITIES

The U.S. rate for accidental deaths in 1980 was 46.2 per 100,000 population. In cities with populations of 200,000 or more, Tulsa, Oklahoma, ranked as the leader in accidental fatalities, with a rate almost twice the national rate. The list of the ten cities with the highest accidental death rates:

1. Tulsa, Okla.		88.6*
2. Oklahoma City, Okla.		67.5
3. New Orleans, La.		65.3
4. Birmingham, Ala.		62.3
5. Kansas City, Mo.		62.0
6. Cleveland, Ohio		51.0
7. Phoenix, Ariz.		50.1
8. Dallas, Tex.		47.9
9. St. Louis, Mo.		47.2
10. Tucson, Ariz.		46.9

*Per 100,000 population.

THE TEN MOST ACCIDENT-PRONE STATES
(accidental deaths per 100,000 population in 1980)

1. Wyoming		85.8
2. Nevada		85.6
3. Montana		75.2
4. New Mexico		72.2
5. Mississippi		71.4
6. Idaho		63.9

7. Arizona	62.2
8. Louisiana	61.1
9. South Dakota	57.8
10. Alabama	57.7

THE TEN COUNTRIES WITH THE MOST FATAL ACCIDENTS

(accident death rate per 100,000 population of those countries reporting to the World Health Organization 1977–79)

1. Hungary	68.8
2. Austria	67.6
3. Venezuela	61.6
4. Luxembourg	61.1
5. Northern Ireland	56.2
6. Poland	54.7
7. West Germany	48.9
8. United States	48.4
9. Switzerland	48.4
10. Canada	48.0

Safety—the Absence of Error

AT WORK—THE SAFEST PLACE TO BE

The most dramatic reduction in accidents has probably occurred in the workplace. Between 1912 and 1980 the accidental death rate per 100,000 workers dropped from 21 to 6—a 71 percent reduction. The significance of this can be seen in one fact: In 1912 approximately 20,000 workers lost their lives in accidents; in 1980, with a work force more than twice as large, only 13,000 accidental work deaths occurred.

The most dangerous industry? Trucking, with a death rate of 36.83 per 100,000 people.

THE SAFEST COMPANY TO WORK FOR . . .

If you want to work where fatal accidents or injuries have been kept out of the workplace, go to work for Du Pont. The four Du Pont plants are in the top twenty listings of places reporting the most hours worked without an occupational injury or illness involving days lost.

THE SAFEST COUNTRY

The safest major country, of those nations reporting to the World Health Organization, is Japan, with an accidental death rate per 100,000 of 26.2—almost half that of the United States.

THE FIVE SAFEST STATES

1. New Jersey	30.2*
2. Connecticut	31.9
3. Maryland	32.6
4. Rhode Island	35.4
5. Delaware	35.5

*Accidental deaths per 100,000 population.

THE SAFEST CITY? HERE'S A SURPRISE

Where do the fewest accidental deaths occur?

In 1980 the safest city in the United States was New York City, with a rate of only 18.7 accidental deaths per 100,000 population. This is far below the U.S. rate of 46.2 deaths per 100,000.

THINK SAFETY; THINK AGAIN

But even the pursuit of safety is no guarantee of freedom from messing up.

During the 1970s the U.S. Consumer Product Safety Commission undertook an admirable safety program. As a measure to help stop needless injuries and deaths to children, it produced 80,000 buttons emblazoned with "Think Toy Safety." The only problem was that the buttons were discovered to have certain faults—such as being coated with lead-based paint, having sharp edges, possessing fasteners that opened too easily, and coming with dangerously pointed ends.

The safety button had to stop being distributed—for safety's sake.

While some errors, especially those that happen to other people, often bring a chuckle or smile to our lips, we have also just seen how errors in the form of accidents have the potential to cause mischief and mayhem. In recent years, however, one dramatic accident demonstrated how a blunder can pose enormous consequences for modern society. The case of Three Mile Island is a disturbing indicator that while sometimes harmless and many times simply frustrating, error also carries with it the potential for catastrophe.

WHILE YOU READ THIS SECTION . . .

During the approximately ten minutes it took you to read this section on accidents, 2 people died from accidents, 190 were injured badly enough to be disabled, and the costs in lost wages, medical care, insurance, and property damage amounted to $1.6 million.

THE WORLD'S MOST EXPENSIVE—AND DANGEROUS—ACCIDENT

It is 3:58 in the morning. March 28, 1979. A high-pitched screech suddenly shatters the calm in the control room of Three Mile Island a massive nuclear generating plant near Harrisburg, Pennsylvania. Hundreds of alarm bells begin ringing. Control gauges go wild. Eight floors below, machinery groans and shudders. The worst nuclear power accident in U.S. history is under way.

A combination of a simple mechanical failure and human error eventually brought Three Mile Island (TMI) to within less than one hour of the uncovering of its nuclear reactor—a development that would have resulted in a catastrophic radioactive contamination of a large area and caused thousands of deaths initially from radiation, with thousands more eventually from cancer.

Although the ultimate tragic outcome was averted, the direct damage to the plant (estimated at $1 billion), combined with the economic loss of the shutdown of the entire facility, resulted in a total cost to the General Public Utilities (GPU) Nuclear Corporation, the plant operator, of $4 billion. As of this writing, more than four years later, Three Mile Island was still too disabled and contaminated to be returned to service.

The major damage to the nuclear facility occurred when plant operators did not know until hours after the accident that the core of the reactor was losing cooling water and dangerously overheating. The operators were judging how much water was in the system by checking the level in a pressurizer, but when the pressurizer indicated a high level of water, they ignored the computer controls and turned off the emergency water injection system out of fear that the system would overfill (the emergency system had been known to malfunction before). What they did not know was that a vital pressure relief valve called a PORV had been stuck open for more than two hours, allowing thousands of gallons of cooling water to escape (the PORV indicator light had gone out in the control room, signaling a closure, but the PORV had not closed).

Thus, instead of too much water in the system, as the plant operators thought, there was too little. With the coolant system failing and the water level falling, the core of the nuclear reactor began boiling dry, and the entire reactor threatened to melt; that would have released tons of radioactivity into the air and ground.

In *The Warning: Accident at Three Mile Island* (W. W. Norton, 1982), authors Mike Gray and Ira Rosen show how these lapses by both men and machine almost led to a nuclear plant catastrophe—a possibility once reserved only for the nightmares of the most rabid pessimists and antinuclear activists.* The authors also show that flaws in the system had been found months before the accident, but in a classic case of Murphy's Law, nothing was done to make certain this information was followed up. "If any one of half a dozen people had read his mail, the accident might never have happened," the authors write.

Indeed, the world's most expensive—and potentially most dangerous—accident was the result of mistakes in the spheres of business, government, and science. An investigation showed that it was not simply operator error or mechanical failure that caused the accident. Rather, it was a combination of factors that bring into sharp focus the potential for error in modern life.

For instance, following investigation of the accident, the manufacturer of TMI, the Babcock & Wilcox Company, was fined more than

---

*Twelve days before the accident at Three Mile Island, *The China Syndrome,* a motion picture forecasting a similar nuclear plant accident, opened in 700 theaters across America. Shortly thereafter—and before the near meltdown at Three Mile Island—*Newsweek* columnist George Will criticized the film and its makers for presenting such a scenario: "To manipulate the audience into antinuclear hysteria, the film suggests that nuclear-power companies carelessly risk destroying their billion-dollar investments. . . . [T]here is more cancer risk in sitting next to a smoker than next to a nuclear plant."

$100,000 for various weaknesses in its design of the plant. The management of Three Mile Island, the General Public Utilities Corporation, was criticized for having been less than diligent in correcting problems quickly (the important closing mechanism, the PORV, had been leaking at TMI for six months prior to the accident).

And even the Nuclear Regulatory Commission (NRC) came in for criticism: It had not been notifying all the nuclear plants of problems encountered in any one of the plants (such a procedure might have alerted TMI personnel to check for stuck-open PORVs since 9 out of 150 PORV openings in U.S. nuclear plants had been found to be faulty).

In fact, in the wake of the costly foul-up at Three Mile Island, General Public Utilities sued both Babcock & Wilcox and the Nuclear Regulatory Commission for failing to warn it of safety hazards. In the spring of 1983 Babcock & Wilcox agreed to a $37 million out-of-court settlement. A $4 billion lawsuit against the NRC was then still pending.

Meanwhile, GPU agreed in April 1983 to pay more than $20 million in cash and stock to shareholders who had bought its shares during the five years prior to the accident. A class action suit, filed in May 1979, had charged that GPU failed to disclose in its prospectuses and reports that the utility might suffer severe financial losses from a nuclear plant accident. In its announcement of the settlement, GPU said that it denied the plaintiff's charges. The extent of damages it had sustained because of Three Mile Island was "totally unforeseen by government and industry."

Three Mile Island is one of seventy-eight operating reactors in the

During the first days of the crisis at Three Mile Island, to relieve tension, people in the Harrisburg area, site of the reactor, swapped jokes about the possible calamitous result of the accident, according to a report in *Psychology Today* (April 1980). Two examples: "Do you know the five-day forecast for Harrisburg? Two days." And "What melts on the ground and not in your mouth? Hershey, Pennsylvania." The jokes seemed to stop, however, when the governor ordered a partial evacuation.

United States. The government's Nuclear Regulatory Commission reviewed the Three Mile Island accident to see what lessons could be learned for the future operation of all of America's reactors. The result was a thick volume and a 347-item action plan to resolve problems and correct mistakes.

And there was much to correct. The president's Commission on the Accident at Three Mile Island, appointed soon after the crisis, found the following lapses at TMI: ambiguous instrument readings, poor control room design, inadequate operator training, inadequate management, lack of accident reporting, inadequate emergency procedures, and malfunctioning hardware. It also found a mismanaged Nuclear Regulatory Commission.

One result of all the studies and recommendations was the assignment of NRC staff to review unusual occurrences in each plant weekly. Heading that operation was a longtime NRC executive, Harold Denton. Interviewed four years after the accident by *Newsweek* magazine ("The Lessons Learned at Three Mile Island," Update section, June 27, 1983), Denton offered the view, based upon his extensive experience with nuclear reactors, that "the real touchstone of reactor safety is the human element and not the hardware."

[The engineers] have discovered once again that the laws of Newton and Einstein are no more immutable than the laws of Murphy.

—Mike Gray and Ira Rosen, *The Warning*

One example: When the "most significant event since Three Mile Island" occurred at New Jersey's Salem 1 plant, where two boron rods controlling the reactor core's nuclear chain reaction failed, new control rods were ordered. But no one inspected the replacements, and workmen installed a broken piece.

"We dream up these technologically sophisticated machines and forget the humans," Denton said.

But it seems that even in the technological dreaming, the human propensity to err intrudes. Stone & Webster, Inc., one of the prime architects of nuclear plants, was later found to have used the wrong computer program to determine whether its plants could withstand earthquakes. Five nuclear plants—all on the East Coast—were affected by this mistake. As a result, the entire mathematical basis for the design of these facilities had to be reexamined.

Welcome to the nuclear age.

Excuse Me:
The Mistake About Mistakes

I make the average number of mistakes.
Maybe 150 or so on a busy day.

—Russell Baker,
Pulitzer Prizewinner,
in his nationally syndicated
newspaper column

The president of the lower house of Austria's parliament, secretly opposed to a pending legislative session, opens parliament by announcing, "I declare this meeting closed."

During a social gathering the Jewish-born wife of a Gentile husband, trying her hardest to blend into high society circles, wants to tell her children to play outside, but instead of using the German word for youngsters (*Jungen*), she calls them *Juden*—Jews.

These two blunders are among many cited and studied by Sigmund Freud, the founder of psychoanalysis, in his major work on human error—*The Psychopathology of Everyday Life,* first published in 1901.

Freud was intrigued with error. In his writings he sought to reveal how slips of the tongue and pen had origins not in chance but in unconscious motivations. While several psychologists before him had begun investigating slips, Freud was the first to set out to explain in a systematic way the nature and origin of human errors, written or spoken. His goal was to show the unconscious to be a factor in the day-to-day events—and misevents—of ordinary life.

The Austrian-born psychiatrist referred to error as *Fehlleistung,* a German word meaning "faulty function" or "blunder." In this category he placed the misreading of words, forgetting of words and names, mislaying of objects, losing things, and making mistakes against one's better judgment. His conclusion about the basis for these mistakes: "When someone makes a slip of the tongue, it is not chance, nor difficulty in articulation either, nor similarity in sound that is responsible; but that in every case a disturbing group of ideas—a complex—can be brought to light which alters the meaning of the intended speech under the guise of an apparent slip of the tongue."

At one point in his book Freud appears to allow room for chance mistakes, but later he states that all errors have their roots in repressed, unconscious causes. In 1907, he wrote, "I have repeatedly been able to show that the most insignificant and obvious errors in speaking have their meaning and can be explained in the same way as the more striking instances."

Thus was born the term *Freudian slip.*

Today many scientists do not agree with Freud that all slips have hidden motives or stem from the unconscious. Current studies trace many mistakes to accidents in the way the brain processes information. Other causes can also be cited—habit, laziness, ineptitude, distractions —leaving a small but hardy band of errors caused by subconscious motivations.

Still, Freud's interest in exploring the origin and nature of error was well founded. As a psychoanalyst and a physician he was concerned with the human being, and errors have been shown to be, above all else, a product of being human.

The study of error making is even more essential in today's high-risk world than it was in Freud's time—or, indeed, in any other time in history. Living as we do amid so much technology, we forgo at our peril a concerted effort to understand, plan for, and, where possible, prevent error.

But we must also be realistic about error. We will never eliminate its existence. Error is democratic and universal, affecting and afflicting us all. Even when we try to design the human being out of a technological system in order to reduce error, we must remember that an error-prone human is designing that system.

Indeed, we will get through life a little more successfully if we adhere to the philosophy of Robert Byrne, the chess grand master. He once wrote in his chess column in the *New York Times*, "It's a mistake not to expect mistakes."

Behind that statement lies the strongest proof of all—thousands of years of human history.

A SAMPLING OF FREUDIAN SLIPS

President Jimmy Carter, at the 1980 Democratic National Convention in a speech honoring former Vice President Hubert H. (for Horatio) Humphrey, referred to him as "Hubert Horatio Hornblower."

Lowell Thomas, the newscaster and world traveler, announced on radio that "this report is credited to the president of the British Board of Trade, Sir Stafford Crapps— Cripps, please."

Dr. Christiaan Barnard, the first surgeon to perform a heart transplant operation, was once introduced to an audience by an emcee who said, "We now have Dr. Christiaan Barnard who will talk about his now famous rear transplant —I mean, rare transplant."

Ralph Nader, the consumer advocate, appeared on the Mike Douglas TV show and announced his new consumer campaign: "We are looking into some of the claims made by a leading booby foob company."

In 1915 the *Washington Post* published a society item that President Woodrow Wilson, then engaged to Edith Galt, had escorted his fiancée to the theater. The *Post* meant to note that instead of paying attention to the play, the president had devoted his attention to entertaining his betrothed. However, because of a typesetter's slip, the *Post* reported that President Wilson, while at the theater, had "spent most of his time entering Mrs. Galt."

A FINAL THOUGHT ABOUT ERROR

In the course of history, one individual stands out as our hero. Whereas Freud analyzed slipups, this person provided a way to correct them.

Here, then, is the profile of the first person to offer humanity help in the never-ending struggle against error:

OUR HERO
The Inventor of the Eraser

JOSEPH PRIESTLEY
(1733–1804)

The eighteenth-century English chemist Joseph Priestley has a special place in the annals of error, for as the inventor of the eraser, he brought a saving grace to all of us who have ever put pencil or pen to paper and then regretted the imperfect result.

The invention of the eraser came to Priestley when he studied the sap of a South American tree newly introduced to Europe. He discovered that the sticky substance from the tree, which the Indians called *caoutchouc,* or "weeping wood," could be used to remove writing. He described this material as "excellently adapted to the purpose of wiping from paper the marks of black lead pencils." Priestley even gave it the name rubber since the removal of a pencil mark resulted from rubbing the hardened sap against the page.

Actually Priestley, whose work broke new ground in chemistry, has several more reasons to be classed as a hero in the history of error: He made some of his most important scientific discoveries through error.

For instance, he was treating oil of vitriol to find out its properties but learned nothing until he accidentally dropped some mercury into the liquid, whereupon "vitriolic acid air," as he called it, evolved. This gas is now called sulfur dioxide.

Priestley also is given credit for the discovery of a colorless gas new to the world—oxygen, although he didn't name it that, misunderstood the basis of its properties, and did not realize the importance of his find. He called the gas dephlogisticated air because he supported the then-current mistaken theory of phlogiston, the hypothetical principle that regarded fire as a material substance. Lavoisier later showed the theory to be incorrect and coined the name oxygen for the gas Priestley had discovered but misunderstood.

NO. # 2

Joseph Priestley • Inventor of the Eraser

Sometimes our hero could be very right indeed. The son of a Nonconformist preacher and himself a Unitarian minister, Priestley supported the American colonists then revolting against George III, opposed the slave trade, and spoke out against religious bigotry. His advocacy of the French Revolution provoked a mob to burn his chapel and sack his house in 1791. A friend of Benjamin Franklin, Priestley left England in 1794 and settled in the United States, where he was warmly welcomed, enjoying the friendship not only of Franklin but also of Thomas Jefferson. He lived the last ten years of his life peacefully in Pennsylvania.

Priestley is also important in the history of error for another discovery of his—a new use for carbon dioxide. He dissolved the gas in water and stumbled across a surprising development: The gas added a delightful taste to the water. He was awarded the Royal Society's Copley Medal for this discovery. Today Priestley's combination of carbon dioxide and water is called seltzer.

Interestingly one of the properties of seltzer is its ability to remove stains, especially stains on clothing or tablecloths caused by blunders at the dinner table.

Thus, Priestley—father of the eraser and of seltzer, two aids for correcting error—was obviously an individual who understood the human condition. Thanks to Priestley and those who have come after him (such as the inventors of Wite-Out, Ko-Rec-Type, and erasable ink), life is a little easier in a world ruled by the Reign of Error, the one earthly reign that will probably never end.

A WORD TO THE READER

You can be part of any sequels to *The Blunder Book*. If you know about or come across an unusual, humorous, or revealing mistake, I would like to hear from you. After all, I have scratched only the surface of error in this book. Journalism, law, art, music, the military—all these subjects and more await their own chapters.

Send your item to me in care of the publisher, William Morrow, 105 Madison Avenue, New York, N.Y. 10016. I cannot answer all correspondence, but entries used will list the name of the contributor.

Be sure to include a reputable source to substantiate the accuracy of your item. And don't forget to proofread your submission. I am no different from anyone else. The only mistakes I tolerate are my own.

ACKNOWLEDGMENTS

It is customary for an author to thank those who contributed in some way to the creation of his book, either by assisting with research, critiquing the manuscript, suggesting changes, or just offering encouragement through countless nights and rewrites. In this regard, I have a number of people to thank for making *The Blunder Book* a reality.

George Gipe, a multitalented writer and friend, was one of the first to encourage me to carry out my idea to write a book on how errors have affected history. Alan Shecter, another good friend, was especially helpful with his reactions to the material as it evolved. My agent, Ellen Levine, was an enthusiastic supporter of the concept from the beginning and offered wise advice and counsel throughout. My editor, Nicholas Bakalar, shepherded the manuscript through its various stages with great care and many invaluable suggestions.

Dr. Arnold Blumberg, professor of history at Towson State University, Dr. Eli Schmell, a biochemist and visiting professor in pharmacology at Johns Hopkins Medical School, and Mark Reches, an engineer with the U.S. Department of Defense, read parts of this book and were of significant help with their clarification about material in their respective fields.

Others who aided in important ways were Jacob Beser, Gail Chalew, Debi Chernak, copy editors Pearl Hanig and Amy Edelman, as well as my devoted wife, Barbara, who once again proved she is this writer's best friend; my children, Aviva, Stuart, and Seth (Stuart was especially helpful with research for the sports chapter), and my brother, Victor. As always, my parents were diligent in bringing items and supporting information to my attention.

I also extend my appreciation to the staffs of the Enoch Pratt Free Library in Baltimore City, the William H. Welch Medical Library of The Johns Hopkins University, and the Baltimore County Library system—all of whom were patient with my questions and persevered in securing the answers.

It is also customary for an author to say that although he owes so much to so many, the fault for any errors that might appear in the finished work lies with nobody but himself. But as, it is hoped, this book has shown, the reason for blunders is not so simple. In a civilized society, we are all forced to rely on others. Consequently, any flubs in this work could be due to a number of causes besides myself: faulty information or mistakes in sources, a secretary's lapse in typing, a goof in typesetting or proofreading.

So don't immediately blame me for the mistakes you may find in this book. I'll accept the blame for some. Which ones, though, I'm not saying.

—M. Hirsh Goldberg

Baltimore, Maryland
April 1984

INDEX